CHARACTERIZATION OF REMOTE-HANDLED TRANSURANIC WASTE FOR THE WASTE ISOLATION PILOT PLANT

FINAL REPORT

Committee on the Characterization of Remote-Handled Transuranic Waste for the Waste
Isolation Pilot Plant
Board on Radioactive Waste Management
Division on Earth and Life Studies
National Research Council

NATIONAL ACADEMY PRESS
Washington, D.C.

NOTICE: The project that is the subject of this report was approved by the Governing Board of the National Research Council, whose members are drawn from the councils of the National Academy of Sciences, the National Academy of Engineering, and the Institute of Medicine. The members of the committee responsible for the report were chosen for their special competence and with regard for appropriate balance. Support for this study was provided by the U.S. Department of Energy under cooperative agreement number DE-FC01-99EW59049. All opinions, findings, conclusions, and recommendations expressed herein are those of the authors and do not necessarily reflect the views of the U.S. Department of Energy.

International Standard Book Number: 0-309-08460-1
Additional copies of this report are available from: National Academy Press 2101 Constitution Avenue, N.W. Box 285 Washington, DC 20055 800–624–6242 202–334–3313 (in the Washington metropolitan area) http://www.nap.edu

Cover. Remote-handled transuranic waste re-packaging operations in a "hot cell" at Battelle Columbus Laboratories-West Jefferson North Campus. From top to bottom: a robotic manipulator arm transfers several pool filters into a new container and closes the lid. In the background: picture of sludge at the bottom of the K-Basins at the Hanford Site. The sludge will be removed and pieces larger than 0.25 inches will be separated. The sludge, which is expected to be remote-handled transuranic waste, will be solidified while the larger pieces will be handled as debris waste. The debris could be transuranic or low-level waste.
Copyright 2002 by the National Academy of Sciences. All rights reserved.

Printed in the United States of America.

THE NATIONAL ACADEMIES

National Academy of Sciences
National Academy of Engineering
Institute of Medicine
National Research Council

The **National Academy of Sciences** is a private, nonprofit, self-perpetuating society of distinguished scholars engaged in scientific and engineering research, dedicated to the furtherance of science and technology and to their use for the general welfare. Upon the authority of the charter granted to it by the Congress in 1863, the Academy has a mandate that requires it to advise the federal government on scientific and technical matters. Dr. Bruce Alberts is president of the National Academy of Sciences.

The **National Academy of Engineering** was established in 1964, under the charter of the National Academy of Sciences, as a parallel organization of outstanding engineers. It is autonomous in its administration and in the selection of its members, sharing with the National Academy of Sciences the responsibility for advising the federal government. The National Academy of Engineering also sponsors engineering programs aimed at meeting national needs, encourages education and research, and recognizes the superior achievements of engineers. Dr. Wm. A. Wulf is president of the National Academy of Engineering.

The **Institute of Medicine** was established in 1970 by the National Academy of Sciences to secure the services of eminent members of appropriate professions in the examination of policy matters pertaining to the health of the public. The Institute acts under the responsibility given to the National Academy of Sciences by its congressional charter to be an adviser to the federal government and, upon its own initiative, to identify issues of medical care, research, and education. Dr. Harvey V. Fineberg is president of the Institute of Medicine.

The **National Research Council** was organized by the National Academy of Sciences in 1916 to associate the broad community of science and technology with the Academy's purposes of furthering knowledge and advising the federal government. Functioning in accordance with general policies determined by the Academy, the Council has become the principal operating agency of both the National Academy of Sciences and the National Academy of Engineering in providing services to the government, the public, and the scientific and engineering communities. The Council is administered jointly by both Academies and the Institute of Medicine. Dr. Bruce Alberts and Dr. Wm. A. Wulf are chairman and vice chairman, respectively, of the National Research Council.

www.national-academies.org

COMMITTEE ON THE CHARACTERIZATION OF REMOTE-HANDLED TRANSURANIC WASTE FOR THE WASTE ISOLATION PILOT PLANT

EULA BINGHAM, *Chair*, University of Cincinnati, Ohio
SANFORD COHEN, S.Cohen & Associates, McLean, Virginia
MILTON LEVENSON, Independent Consultant, Menlo Park, California
KENNETH L.MOSSMAN, Arizona State University, Tempe
ERNEST NIESCHMIDT, Idaho State University, Idaho Falls
JOHN PLODINEC, Mississippi State University, Starkville
ANNE E.SMITH, Charles River Associates, Washington, D.C.

Consultant

HEINO NITSCHE, University of California, Berkeley

Liaison to the Board on Radioactive Waste Management

ALEXANDER MACLACHLAN, E.I.du Pont de Nemours and Company (retired), Wilmington, Delaware

Staff

BARBARA PASTINA, Study Director
ANGELA R.TAYLOR, Senior Project Assistant
DARLA J.THOMPSON, Research Assistant

BOARD ON RADIOACTIVE WASTE MANAGEMENT

JOHN F.AHEARNE, *Chair,* Sigma Xi and Duke University, Research Triangle Park, North Carolina
CHARLES MCCOMBIE, *Vice Chair,* Consultant, Gipf-Oberfrick, Switzerland
ROBERT M.BERNERO, U.S. Nuclear Regulatory Commission (retired), Gaithersburg, Maryland
ROBERT J.BUDNITZ, Future Resources Associates, Inc., Berkeley, California
GREGORY R.CHOPPIN, Florida State University, Tallahassee
RODNEY EWING, University of Michigan, Ann Arbor
JAMES H.JOHNSON, JR., Howard University, Washington, D.C.
HOWARD C.KUNREUTHER, University of Pennsylvania
NIKOLAY LAVEROV, Russian Academy of Sciences, Moscow
MILTON LEVENSON, Bechtel International (retired), Menlo Park, California
JANE C.S.LONG, Mackay School of Mines, University of Nevada, Reno
ALEXANDER MACLACHLAN, E.I.du Pont de Nemours and Company (retired), Wilmington, Delaware
NORINE E.NOONAN, College of Charleston, South Carolina
EUGENE A.ROSA, Washington State University, Pullman
ATSUYUKI SUZUKI, University of Tokyo, Japan
VICTORIA J.TSCHINKEL, Victoria J.Tschinkel Environmental Policy and Management, Tallahassee, Florida

Staff

KEVIN D.CROWLEY, Director
MICAH D.LOWENTHAL, Staff Officer
BARBARA PASTINA, Senior Staff Officer
JOHN R.WILEY, Senior Staff Officer
TONI GREENLEAF, Administrative Associate
DARLA J.THOMPSON, Research Assistant
LATRICIA C.BAILEY, Senior Project Assistant
LAURA D.LLANOS, Senior Project Assistant
ANGELA R.TAYLOR, Senior Project Assistant
JAMES YATES, JR., Office Assistant

Preface

The U.S. Department of Energy (DOE) is seeking authorization to dispose of its remote-handled transuranic (RH-TRU) waste in the Waste Isolation Pilot Plant (WIPP), a geologic repository located in New Mexico. The DOE-Carlsbad Field Office asked the National Research Council to provide an independent review of its proposed characterization plan for RH-TRU waste.

To this end, the National Research Council appointed an ad hoc committee of seven members and one consultant with expertise in the following disciplines: knowledge of the DOE weapons complex, particularly with respect to the generation of RH-TRU waste; TRU waste characterization techniques; health physics; actinide chemistry; environmental policy; public policy; and risk assessment. The committee's statement of task is in Sidebar P.1. The biographical sketches of committee members can be found in Appendix A.

SIDEBAR P.1 STATEMENT OF TASK

The objective of this study is to review and evaluate the U.S. Department of Energy's (DOE's) plan to characterize remote-handled transuranic (RH-TRU) waste to be disposed of at Waste Isolation Pilot Plant (WIPP). The committee will provide recommendations, as necessary, for improving the plan's technical soundness, protection of worker safety, and compliance with applicable regulatory requirements. This study does not address transportation issues related to RH-TRU waste.

Examples of criteria that could be used by the committee to review DOE's characterization plan for RH-TRU waste are the following: validity of assumptions used for the proposed plan; quality of information available on RH-TRU waste; uncertainties on process knowledge data and consequences if wrong; appropriateness and sensitivity of overall system safety to uncertainties in the waste specifications; limitations of the methodologies and procedures to characterize RH-TRU waste; and validity of DOE's conclusions concerning its characterization plan.

The committee met four times, from June 2001 to May 2002, to complete its review of DOE's characterization plan. The committee interacted with DOE, generator sites, WIPP's primary regulatory agencies (Environmental Protection Agency and the New Mexico Environment Department), the New Mexico Environmental Evaluation Group, and the public to gather information on the status of RH-TRU waste across DOE's complex, characterization requirements, and expectations for this new plan. The diverse mix of participants from DOE (headquarters and the sites), regulatory agencies, and national laboratories contributed to lively discussions and great insights into the committee's task.

The committee gathered information from the following sites: Battelle Columbus Laboratories, Idaho Engineering and Environmental Laboratory, Oak Ridge National Laboratory, Los Alamos National Laboratory, Argonne National Laboratory-West, and

PREFACE

the Hanford Site. These sites were selected because they are the main RH-TRU waste generators or, in the case of Battelle Columbus Laboratories, because this site has begun RH-TRU waste characterization operations. The committee also visited Battelle and the WIPP site. Information-gathering meeting agendas and speakers are listed in Appendix B.

This report has been reviewed in draft form by individuals chosen for their diverse perspectives and technical expertise, in accordance with procedures approved by the National Research Council's Report Review Committee. The purpose of this independent review is to provide candid and critical comments that will assist the National Research Council in making the published report technically sound and to ensure that the report meets National Research Council institutional standards for objectivity, evidence, and responsiveness to the study charge. The review comments and draft manuscript remain confidential to protect the integrity of the deliberative process. We wish to thank the following individuals for their participation in the review of this report:

Robert M.Bernero, U.S. Nuclear Regulatory Commission (retired)

David C.Camp, Lawrence Livermore National Laboratory

Rodney C.Ewing, University of Michigan

Harry D.Harmon, Harmon Consulting

Darleane C.Hoffman, Lawrence Berkeley Laboratory

Robert H.Neill, Environmental Evaluation Group (retired)

Victoria J.Tschinkel, Victoria J.Tschinkel Environmental Policy and Management

Raymond G.Wymer, Oak Ridge National Laboratory (retired)

Although the reviewers listed above have provided many constructive comments and suggestions, they were not asked to endorse the conclusions or recommendations, nor did they see the final draft of the report before its release. The review of this report was overseen by Chris G.Whipple, ENVIRON International, Inc. Appointed by the National Research Council, he was responsible for making certain that an independent examination of this report was carried out in accordance with institutional procedures and that all review comments were carefully considered. Responsibility for the final content of this report rests entirely with the authoring committee and the institution.

This study could not have been completed without the assistance of many individuals and organizations. The committee thanks many DOE staff members in the Carlsbad Field Office, in the Office of Environmental Management, and at RH-TRU waste generator sites, including contractors, for their active participation in committee meetings and for promptly responding to requests of information. The committee is especially grateful to Roger Nelson, chief scientist at the DOE-Carlsbad Field Office, who served as primary contact for this study and provided outstanding support to committee's activities; Clayton Gist, RH-TRU waste program manager, and Inés Triay, manager of the Carlsbad Field Office. The committee also expresses its deep appreciation to those who organized site tours, especially James Eide at Battelle Columbus Laboratories and Norbert Rempe at the WIPP.

The committee is equally grateful to Steve Zappe of the New Mexico Environment Department and Scott Monroe of the U.S. Environmental Protection Agency for their input on WIPP's regulatory requirements. Matthew Silva, George Anastas, and James Channell of the New Mexico Environmental Evaluation Group also provided the committee with helpful technical and historical perspectives on the characterization of RH-TRU waste.

Finally, the committee thanks the following National Research Council staff members: Kevin Crowley, who provided guidance in separating science from policy; Angela Taylor, who made certain our committee meetings were as pleasant as possible; Darla Thompson who provided strong research support; and Barbara Pastina, whose knowledge, patience, and organizational and writing skills are unparalleled. She anticipated the obstacles and smoothed the way for the committee to deal with issues in the statement of task during this fast-track project. Any success of the project is because of her dedication.

Eula Bingham, *Chair*
Committee on the Characterization of Remote-Handled Transuranic Waste for the WIPP
July 2002

PREFACE

Contents

	Executive Summary,	*1*
1	Introduction,	*8*
2	Remote-Handled Transuranic Waste,	*15*
3	Regulatory Context for the Disposal of Remote-Handled Transuranic Waste,	*27*
4	Department of Energy's Proposed Characterization Plan,	*34*
5	Assessment of the Proposed Characterization Plan,	*47*
	References,	*62*
	Appendixes	
A	Biographical Sketches of Committee Members,	*65*
B	Information-Gathering Meetings,	*68*
C	Excerpt from the Committee's Interim Report: Chapter 5,	*71*
D	DOE's Response to the Committee's Interim Report,	*79*
E	Information About Selected Transuranic Waste Generator Sites,	*91*
F	Overview of the Contact-Handled Transuranic Waste Characterization Plan,	*103*
G	Non-Destructive Techniques for Remote-Handled Transuranic Waste Characterization,	*105*
H	Waste Dose Rates and Characterization Cost Estimates,	*109*
I	Glossary,	*112*
J	Acronyms,	*120*

CONTENTS

Executive Summary

The Department of Energy's (DOE's) Carlsbad Field Office asked the National Research Council to review a proposed characterization plan for remote-handled transuranic (RH-TRU)[1] waste and to provide recommendations, as necessary, for improving the plan's technical soundness, protection of worker safety and health, and compliance with regulatory requirements.

There are approximately 3,800 cubic meters of defense-related RH-TRU waste to be removed from DOE's weapons complex. DOE is seeking authorization to dispose of this RH-TRU waste in the Waste Isolation Pilot Plant (WIPP), a geologic repository in southeastern New Mexico. WIPP is regulated by the Environmental Protection Agency (EPA) and the New Mexico Environment Department (NMED).

WIPP is currently certified by the EPA and permitted by the NMED to dispose only of contact-handled (CH) and mixed[2] CH-TRU waste. To obtain authorization to ship RH-TRU waste to WIPP, DOE must submit to the regulatory agencies a characterization plan for RH-TRU waste and request a modification of the EPA Certification and RCRA Permit. The regulatory agencies may accept, reject, or propose modifications to the characterization plan and, eventually, may authorize RH-TRU waste disposal in WIPP.

S.1 TECHNICAL AND REGULATORY CONTEXT OF RH-TRU WASTE CHARACTERIZATION

Remote-handled TRU waste presents a long-term (i.e., 10,000 years) radiological hazard associated with TRU radionuclides (related to the presence of long-lived alpha-emitting isotopes) and a short-term (less than 300 years) radiological hazard associated with short-lived gamma-emitting radionuclides (related to the presence of fission and activation products). Gamma radiation can penetrate through the walls of waste containers and the skin; therefore, RH-TRU waste presents a potential for radiation doses to workers during waste characterization, handling, and emplacement. The associated costs due to the heavy shielding and remote-handling equipment, necessary to protect workers from radiation, are also significant. The long-term radiological hazard of RH-TRU waste is associated with the potential for TRU radionuclide release into the environment once the waste is disposed of in WIPP.

S.1.1 DOE's RH-TRU Waste Inventories

While RH-TRU waste volume represents a small fraction of the allowed TRU waste in WIPP (3,800 cubic meters out of approximately 176,000 cubic meters), it accounts for

---

[1] Transuranic waste is radioactive waste containing alpha-emitting radionuclides of atomic number greater than 92, half-life greater than 20 years, and activity greater than 100 nanocuries per gram of waste. Transuranic waste is classified as remote-handled or contact-handled depending on the radiation dose rate at the surface of the package. A more detailed definition is given in Sidebar 1.1.

[2] Transuranic waste can be mixed with hazardous components regulated under the Resources Conservation and Recovery Act (RCRA). This type of waste is called mixed TRU waste and is regulated by NMED as hazardous waste.

14 percent of the total curie activity in DOE's TRU inventory. However, the number of TRU (i.e., from alpha-emitting radionuclides with a half-life greater than 20 years) curies in the RH-TRU waste inventory is 2 orders of magnitude lower than that in the CH-TRU waste inventory expected in WIPP. Also, most of the total (CH plus RH) long-term (i.e., TRU) activity expected in WIPP comes from CH-TRU waste, with only a 0.5 percent contribution from RH-TRU waste (see Table 2.3 in Chapter 2).

Remote-handled TRU waste to be emplaced in WIPP is stored or is to be generated as the result of cleanup activities at 13 DOE weapons complex sites (see Chapter 2). Approximately 1,600 cubic meters of projected RH-TRU waste is waste that originates from the dismantlement of hot cells and other facilities used for defense-related purposes, waste that must be recovered and separated from other wastes (such as at the Hanford Site), or waste that must be reprocessed to eliminate hazardous components prohibited in WIPP (such as at Oak Ridge National Laboratory). Already stored at DOE sites are approximately 2,200 cubic meters of RH-TRU waste, but most of this inventory must be characterized and repackaged to meet waste transportation criteria. Only 5 percent of the waste has been characterized and packaged in a form that could be suitable for shipment to WIPP. To obtain authorization to ship RH-TRU waste to WIPP, generator sites must characterize all their RH-TRU waste inventories according to a plan approved by EPA and NMED.

S.1.2 Regulatory Context

Congress designated WIPP as the nation's defense-related transuranic waste repository in the Land Withdrawal Act of 1992. This Act allows the geologic disposal in WIPP of up to 175,564 cubic meters of TRU waste, including 7,080 cubic meters (4 percent of the total volume) of RH-TRU waste. The EPA, under this Act, regulates the radiological release limits for WIPP over a period of 10,000 years. The NMED, under RCRA, regulates the non-radiological risks related to hazardous waste disposal during the operational phase and up to 30 years following repository closure.

The EPA Certification and RCRA Permit establish characterization requirements for all TRU waste (CH and RH) to be emplaced in WIPP. Contact-handled TRU waste is characterized according to a plan that was negotiated for almost two decades among DOE and the regulators prior to the certification of the WIPP facility. This plan consists of 100 percent confirmation of acceptable knowledge (AK)[3] information through visual examination, radioassay, headspace gas sampling, or radiography. A DOE review of the CH-TRU waste characterization procedures revealed that DOE developed self-imposed waste restrictions in the waste acceptance criteria and in the implementation documents used by generator sites to characterize CH-TRU waste. A previous National Research Council committee found that these self-imposed procedures lack technical, safety, or legal basis (see Sidebar 3.1). DOE is now facing the challenge of proposing a characterization plan for RH-TRU waste that meets EPA and NMED requirements but follows a different approach than that used for CH-TRU waste.

---

[3]According to the EPA's definition, AK consists of historical information on a particular waste stream, which may include administrative, procurement, and quality control documentation associated with the generating process, past sampling and analytical data, information about the process used to generate the waste, material inputs to the process, and the time period during which the waste was generated (see Sidebar 2.1).

S.2 DOE'S PROPOSED CHARACTERIZATION PLAN FOR RH-TRU WASTE

The following two documents include DOE's proposed characterization plan for RH-TRU waste:

- Document 1: Notification of Proposed Change to the EPA Title 40 CFR Part 194 Certification of the WIPP.
- Document 2: Request for RCRA Class 3 Permit Modification to the NMED.

On June 28, 2002 DOE submitted Documents 1 and 2 to EPA and NMED, respectively. The committee reviewed two drafts of the above documents prior to their submission. Findings and recommendations in this report apply to the characterization plan as presented in the March 2002 draft.

DOE's goal is to adopt a performance-based approach to meet EPA and NMED requirements for waste characterization. This approach relies on the impact of the RH-TRU waste inventory on the performance of WIPP. The Sandia National Laboratories undertook two performance assessment analyses showing that the characteristics of RH-TRU waste will have a negligible impact on the potential long-term release of radionuclides into the environment. This is due to the characteristics of DOE's RH-TRU waste inventories in terms of volume and radiological composition.

To address the EPA and NMED regulatory requirements, DOE proposes to use the following characterization methods: dose-to-curie conversion, visual examination, radiography, direct assay, counting containers, AK, and characterization at the time of packaging. The dose-to-curie method consists of correlating a surface dose rate measurement with documented waste isotopic distributions through the use of empirically developed conversion factors. Visual examination involves the removal of items from the container (by remote-handled methods) for inspection and identification. Visual examination does not provide information on the isotopic composition of waste. Radiography utilizes penetrating radiation (typically X-rays) to investigate the contents of containers. Direct assay consists of radiochemical analyses using the same methods approved for CH-TRU waste characterization. Counting containers is the method used to determine the amount of metal emplaced in WIPP by an automatic inventory of the number of containers during packaging. Characterization at the time of packaging consists of visual examination and the use of other characterization methods (i.e., AK, radiochemical analysis, or radiography) as waste is generated, packaged, or repackaged in a form suitable for shipment to WIPP. The AK method, which is the primary method of compliance for waste characterization, consists of collecting historical information and new information obtained through the "characterization at the time of packaging" method (see Finding 1B below). Further details are provided in Chapter 4.

Unlike the current CH-TRU waste characterization approach, which requires 100 percent confirmation of the AK information, for RH-TRU waste characterization DOE proposes performing confirmatory measurements only on 10 percent of the waste. This confirmation activity is proposed when AK is insufficient to address a characterization requirement.

S.3 COMMITTEE'S ASSESSMENT OF DOE'S PROPOSED CHARACTERIZATION PLAN

This report is not meant to be a comprehensive review of the entire RH-TRU waste program, which encompasses RH-TRU waste transportation, storage, waste certification, operational safety issues, occupational health and safety regulations, waste generator states' regulations, waste acceptance criteria, as well as DOE orders.

The committee used the criteria listed in the statement of task to assess DOE's proposed characterization plan (see Sidebar P.1 in the Preface). Findings and recommendations address: 1) the context of RH-TRU waste characterization, 2) the characterization plan's technical soundness, 3) protection of worker safety and health, and 4) compliance with regulatory requirements. Findings are referenced (in parentheses) according to the order they are presented in Chapter 5. Each finding is followed by the corresponding recommendation (in bold characters). The rationales are provided in Chapter 5.

1) Context of RH-TRU Waste Characterization

The committee did not verify the data provided by DOE on its RH-TRU waste inventories. Acknowledging the past fluctuations of RH-TRU waste inventories, the committee assumes that DOE presented the most up-to-date information available. Even if the information presented in DOE inventories were not accurate, by law the RH-TRU waste emplaced in WIPP cannot exceed 4 percent of the total volume inventory and 5.1 million curies. According to DOE inventories, 95 percent of the RH-TRU waste to be disposed of in WIPP has yet to be generated or needs to be processed, packaged, or repackaged. This waste will be characterized (through visual examination and physical and chemical analyses) at the time of packaging (Finding 1A). **DOE should emphasize the argument that the characterization information collected for most of RH-TRU waste does not need confirmatory measurements because the repackaging or generation of waste will be carried out under a certified quality assurance program.** If the volume of RH-TRU waste represents between 2 and 4 percent of the volume of TRU waste, and the information collected for over 95 percent of RH-TRU waste does not need confirmation, then only the remaining 5 percent of the RH-TRU waste inventory (between 0.1 and 0.2 percent of the total inventory) needs confirmatory activities.

As previously mentioned, DOE uses the term "AK" to indicate both the historical information and the newly generated characterization information collected at the time of waste generation, packaging, or repackaging. However, for 95 percent of the RH-TRU waste inventory, AK refers mostly to the latter (Finding 1B). **The committee recommends that DOE use a different term than "AK" for this newly generated information.** Using AK for both historical and newly generated information is potentially confusing because AK is generally associated with historical information, which requires some type of confirmation.

2) Characterization Plan's Technical Soundness

The committee found that DOE's proposed characterization plan is not completely performance based and that several characterization activities are based on non-technical considerations (Finding 2A). The committee questions the technical basis of some of these characterization activities. The committee acknowledges that non-technical considerations may be important for maintaining effective working relationships among DOE, EPA, and NMED; however, **DOE should propose only characterization activities that have a technical, health and safety, or regulatory basis.**

The committee provides the following examples of activities lacking a technical basis in the context of RH-TRU waste characterization: determination of radiological activity, free water/liquid, ferrous metal, cellulosics, plastic, rubber, and polychlorinated biphenyls content. According to the Sandia impact analyses, these RH-TRU waste parameters will have only a negligible impact on the performance of the repository. Also, DOE did not provide the technical basis for selecting 10 percent of the waste as a representative waste sample to perform confirmation activities.

In some instances, the plan lacked specificity because most of the operational details are site-specific and were not available at the time of writing (Finding 2B). It is the committee's understanding that, along with Documents 1 and 2, DOE will submit three site-specific documents containing a completed characterization plan for selected RH-TRU generator sites. The committee supports this initiative. **The site-specific documents should present clear and technically defensible data qualification requirements for RH-TRU waste characterization.**

Also, tolerable decision error rates are never clearly defined in Documents 1 and 2 (Finding 2C). Tolerable decision error rates are determined by weighing the consequences of mischaracterization against the costs (including worker risks) of achieving better characterization. Once established, tolerable decision error rates can be used to identify the attendant quality assurance requirements for sampling (i.e., the measurement quality objectives, which DOE calls quality assurance objectives[4]). **DOE's proposed characterization plan should address tolerable decision error rates associated with characterization information. These errors should not be overly stringent so as to negatively impact the sites' ability to implement ALARA.**[5]

For the 5 percent of the RH-TRU waste inventory that does not require packaging or repackaging, it is not clear how visual examination and radiography can confirm AK information for prohibited items (Finding 2D). For example, visual examination and radiography cannot distinguish between corrosive and non-corrosive liquids, whereas AK may provide records of the existence of such liquids in the waste. Historical AK may be a better indicator of some of the currently prohibited items than visual examination and radiography. **The characterization plan should clarify under which conditions confirmation of historical AK is warranted and what are the most effective methods proposed.** The committee could not determine the effectiveness of specific technologies, such as X-ray radiography, in providing confirmatory data for the high-dose-rate fraction of RH-TRU waste containers (Finding 2E). **DOE should provide justification for the technologies proposed for obtaining confirmatory data and provide evidence of their effectiveness across the entire spectrum of dose rates for RH-TRU waste.**

Overall, in the context of RH-TRU waste characterization and from a performance point of view, the committee found that the general approach DOE is proposing is technically sound. However, Documents 1 and 2 do not present a performance-based plan as effectively as they could.

3) Protection of Worker Safety and Health

Potential worker radiation doses and related characterization costs are distinguishing features of RH-TRU waste compared to CH-TRU waste, but estimates of worker doses and characterization costs for RH-TRU waste are limited, site-specific, and not completely reliable (Finding 3A). **DOE could strengthen the rationale of its**

---

[4]DOE's "quality assurance objectives" are the accuracy, precision, completeness, comparability, and representativeness of the characterization data.

[5]The ALARA (As Low As is Reasonably Achievable) principle requires that a reasonable effort be made to keep workers radiation exposures as far below the regulatory dose limits as is practical (see Chapter 4).

characterization plan for RH-TRU waste by including a discussion of estimates of worker doses and characterization costs in the three site-specific plans accompanying the submittal documents. Discussion of worker doses and costs is relevant since radiation protection standards and criteria are predicated on the concept of assessing the decrement in risk per increment in cost. It is also important to recognize that tolerable decision error rates associated with characterization methods will have an impact on the implementation of the ALARA principle at generator sites (see Finding 2C).

Documents 1 and 2 provide flexibility to the generator sites in the implementation of the RH-TRU waste characterization plan (Finding 3B). **DOE should continue its effort in ensuring sufficient flexibility to generator sites in the implementation of the characterization plan. However, characterization activities that share common elements across sites should be standardized.** Given the differences among generator sites (for instance, in the composition of waste streams, quality of AK and inventories, and characterization and repackaging facilities), flexibility in a waste characterization plan is important to adapt the sites' implementation programs. A rigid, overly prescriptive characterization plan may lead to unnecessary radiation doses to workers and characterization costs. However, standardization of common elements of the characterization program may facilitate characterization compliance verifications and, possibly, reduce characterization costs.

4) Compliance with Regulatory Requirements

The difficulty of this latter part of the statement of task lies in performing a review of a characterization plan for RH-TRU waste against compliance with existing characterization requirements that currently exclude RH-TRU waste. The task is further complicated by the existence of the characterization approach approved by EPA and NMED that is currently used for CH-TRU waste. It is important to emphasize the difference between regulatory requirements and the approach to meet such requirements. While the regulatory requirements for CH- and RH-TRU waste characterization are the same, the approach to address these requirements can be different. The committee evaluated, from a technical point of view, the approach DOE is proposing to characterize RH-TRU waste and how it addresses regulatory requirements. The committee was not asked to comment on these requirements nor was it asked to determine if the plan complies with the regulatory requirements. The latter is obviously a policy decision belonging to the regulatory agencies.

The committee found that DOE's proposed characterization plan for RH-TRU waste deliberately tracks the characterization plan for CH-TRU waste (Finding 4A). **The committee recommends that DOE evaluate whether existing characterization practices for CH-TRU waste, when applied to the characterization of RH-TRU waste, have an impact on the protection of the environment, health and safety of public and workers, and cost-effectiveness of the characterization program.** Also, like in the CH-TRU waste characterization plan, Documents 1 and 2 include, as basis for characterization objectives, requirements other than those applicable to EPA certification and RCRA permit, for instance transportation requirements or waste acceptance criteria (Finding 4B). **The committee recommends that submittal documents focus on regulatory requirements under the relevant agency's purview and distinguish between these requirements and ancillary information describing the context of RH-TRU waste characterization.** While it is important to provide, in the submittal documents, the full context of the characterization of RH-TRU waste, including material outside the regulatory jurisdiction of EPA and NMED may unnecessarily complicate the regulatory review process. It would be helpful for both DOE and the regulatory agencies if the submittal documents indicate which requirements are to be reviewed and which are given for information purposes. This would allow the regulator and DOE to keep track of various requirements and to streamline characterization practices as experience is gained.

S.3.1 Overall Assessment of DOE's Characterization Plan

The committee observed a net improvement between the July 2001 and March 2002 drafts of the characterization plan. Concerning the plan's technical soundness, DOE itself recognized that this plan is not completely performance based and that other considerations played a role in its development. The committee identified some characterization activities lacking technical bases in the context of RH-TRU waste and a potential technical problem with radiographic examination of waste. Also, DOE's proposed characterization plan does not adequately address the issue of tolerable decision error rates associated with all characterization information.

In some instances, the plan lacks specificity because most of the operational details are site-specific and were not available at the time of writing. The site-specific accompanying documents should provide useful clarifications. In the context of RH-TRU waste characterization and from a performance point of view, the committee found that the general approach DOE is proposing is technically sound. However, Documents 1 and 2 do not present a performance-based plan as effectively as they could.

Concerning the plan's protection of worker health and safety, the committee recommends that the approved characterization plan not include overly stringent tolerable decision error rates that could negatively impact the sites' ability to manage worker risks. It is important to recognize that the allowable uncertainties in the final characterization plan approved by EPA and NMED may have an impact on generator sites' radiation protection programs.

Concerning compliance with regulatory requirements in the EPA Certification and RCRA Permit, the committee did not observe any requirement that was not addressed in DOE's characterization plan. In fact, the proposed characterization plan for RH-TRU waste addresses some requirements that are not under the relevant agency's purview. Moreover, the characterization plan for RH-TRU waste deliberately tracks that for CH-TRU waste. The committee recommends evaluating whether existing characterization practices for CH-TRU waste, when applied to the characterization of RH-TRU waste, have an impact on the protection of the environment, health and safety of public and workers, and cost-effectiveness of the characterization program.

1

Introduction

The purpose of this report is to provide an independent technical review of the U.S. Department of Energy's (DOE's) proposed characterization plan for defense-related remote-handled (RH) transuranic (TRU) waste. Remote-handled transuranic waste is defined in Sidebar 1.1.

Transuranic waste is classified as remote-handled (RH) or contact-handled (CH) depending on the radiation dose rate measured at the surface of waste containers. While CH-TRU waste can be safely handled by direct contact, RH-TRU waste requires heavy container shielding or remote-handling equipment. Therefore, the main issues with RH-TRU waste characterization are the potential for worker exposure to radiation and the associated costs.

Because of its radiological hazard, TRU waste requires geologic isolation. Congress designated the Waste Isolation Pilot Plant (WIPP), in southeastern New Mexico, as the nation's geologic disposal facility for defense-related transuranic waste (U.S. Congress, 1992). Figures 1.1, 1.2, and Sidebar 1.2 provide a schematic representation and a brief description of the WIPP facility. DOE is seeking authorization to dispose of RH-TRU waste in WIPP as part of its commitment to safely manage and clean up contamination from over 50 years of nuclear materials research and production. About 3,800 cubic meters of defense-related RH-TRU waste (corresponding to approximately 1 million curies) is the current estimated volume of RH-TRU waste to be disposed in WIPP.

The U.S. Environmental Protection Agency (EPA) and the New Mexico Environment Department (NMED) are WIPP's regulatory agencies. In 1998, the EPA, which regulates TRU waste, granted WIPP a Certification of Compliance (referred to as the EPA Certification) with the geologic disposal standards established in Title 40 of the Code of Federal Regulations Part 191 (40 CFR 191). In 1999, the NMED, which regulates hazardous waste disposal under the Resources Conservation and Recovery Act (RCRA), granted WIPP a hazardous waste facility permit (referred to as the RCRA Permit). Regulatory requirements addressing radioactive and hazardous waste will be discussed in Chapter 3.

EPA and NMED authorize only CH- and mixed CH-TRU waste in WIPP. Remote-handled TRU waste is not allowed in this facility because, according to these agencies, DOE does not have an appropriate characterization plan for this type of waste (see Chapter 3).

The EPA Certification and RCRA Permit contain characterization requirements for TRU waste to be emplaced in WIPP. These characterization requirements were negotiated among DOE, EPA, and NMED and they are the basis of the current characterization plan for CH-TRU waste.

SIDEBAR 1.1 DEFINITION OF TRANSURANIC WASTE AND CLASSIFICATION

Transuranic (TRU) waste is radioactive waste containing alpha-emitting radionuclides of atomic number greater than 92, half-life[a] greater than 20 years, and activity greater than 100 nanocuries per gram of waste, except for: high-level radioactive waste; waste that the Department of Energy (DOE) has determined, with the concurrence of the Environmental Protection Agency (EPA), does not need the degree of isolation required by the disposal regulations; or waste that the Nuclear Regulatory Commission has approved for disposal on a case-by-case basis in accordance with Title 10 of the Code of Federal Regulations Part 61 (U.S. Congress, 1992).

Transuranic waste can consist of sludge or solid material, such as pieces of clothing, tools, and debris. This type of waste was produced during the processing of nuclear materials and continues to be generated in the cleanup of DOE weapons sites. The main alpha-emitting radionuclides in TRU waste (and their respective half-lives indicated in parenthesis) are plutonium-238 (87.7 years), plutonium-239 (24,100 years), plutonium-240 (6,560 years), and americium-241 (433 years). Other long half-life isotopes, such as uranium-233 (159,200 years), uranium-234 (24,600 years), uranium-235 (704 million years), and uranium-238 (4.47 billion years) may also be present in transuranic (TRU) waste.

Transuranic waste is classified as *contact-handled* (CH) or *remote-handled* (RH), according to the dose rate measured at the container surface. According to Land Withdrawal Act, "the term 'contact-handled transuranic waste' means transuranic waste with a surface dose rate not greater than 200 millirem per hour. The term 'remote-handled transuranic waste' means transuranic waste with a surface dose rate of 200 millirem per hour or greater" (U.S. Congress, 1992). The legal definitions of CH-TRU and RH-TRU waste do not clearly address container surface dose rates of exactly 200 mrem per hour. Therefore, waste packages approaching 200 mrem per hour are handled directly or remotely, depending on site-specific practices.

Transuranic waste is further classified as *mixed* or *non-mixed*. Mixed TRU waste contains both radioactive material regulated under the Atomic Energy Act and hazardous waste material regulated under the Resource Conservation and Recovery Act, or RCRA (U.S. Congress, 1976). RCRA is a federal law addressing hazardous waste designed to ensure that the generation, transportation, treatment, storage, and disposal of hazardous wastes are conducted in a manner that protects human health and the environment (EPA, 1994). The statutory definition of hazardous waste is provided in Section 1004(5) of RCRA as follows (EPA, 1994; page 1–3):

"A solid waste, or combination of solid waste, which because of its quantity, concentration, or physical, chemical, or infectious characteristics may 1) cause, or significantly contribute to an increase in mortality or an increase in serious irreversible, or incapacitating reversible, illness; or 2) pose a substantial present or potential hazard to human health or the environment when improperly treated, stored, transported, or disposed of, or otherwise managed."

Examples of hazardous waste material are ignitable, corrosive, reactive, and toxic substances. For an overview of the Atomic Energy Act and major environmental laws, including RCRA, the reader may refer to "The Nuclear Waste Primer" (The League of Women Voters, 1993).

---

[a] For a definition of half-life see the glossary, Appendix I.

INTRODUCTION

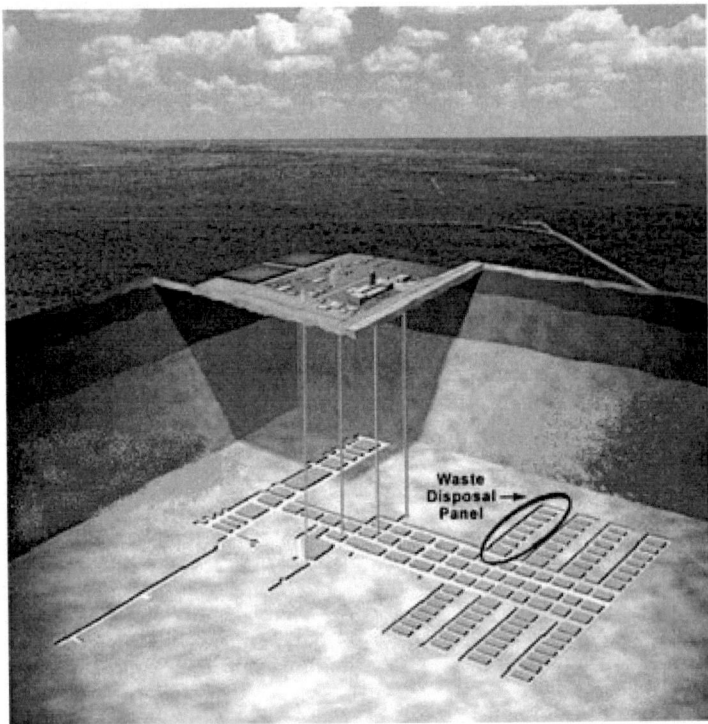

FIGURE 1.1 Schematic representation of the WIPP Facility. The facility will have eight disposal panels, each divided into seven rooms. Only two of the eight panels have been mined to date. SOURCE: DOE (DOE-CBFO, 2000).

To obtain authorization to dispose of RH-TRU waste in WIPP, DOE must request a modification of the EPA Certification and RCRA Permit and present a proposed characterization plan for this type of waste. This proposed characterization plan is the focus of this report.

DOE's proposed characterization plan for RH-TRU waste consists of the following two documents:

- Document 1: Notification of Proposed Change to the EPA Title 40 CFR Part 194 Certification of the WIPP.
- Document 2: Request for RCRA Class 3 Permit Modification[1] to the NMED.

Documents 1 and 2 will often be referred to in this report as the "submittal documents."

---

[1] Class 3 permit modifications, as determined by NMED, are considered major changes to the permit. This class of permit modification requires a public notice, a 60-day comment period, and hearings, including testimony and cross-examination of witnesses, before NMED issues the final draft permit language.

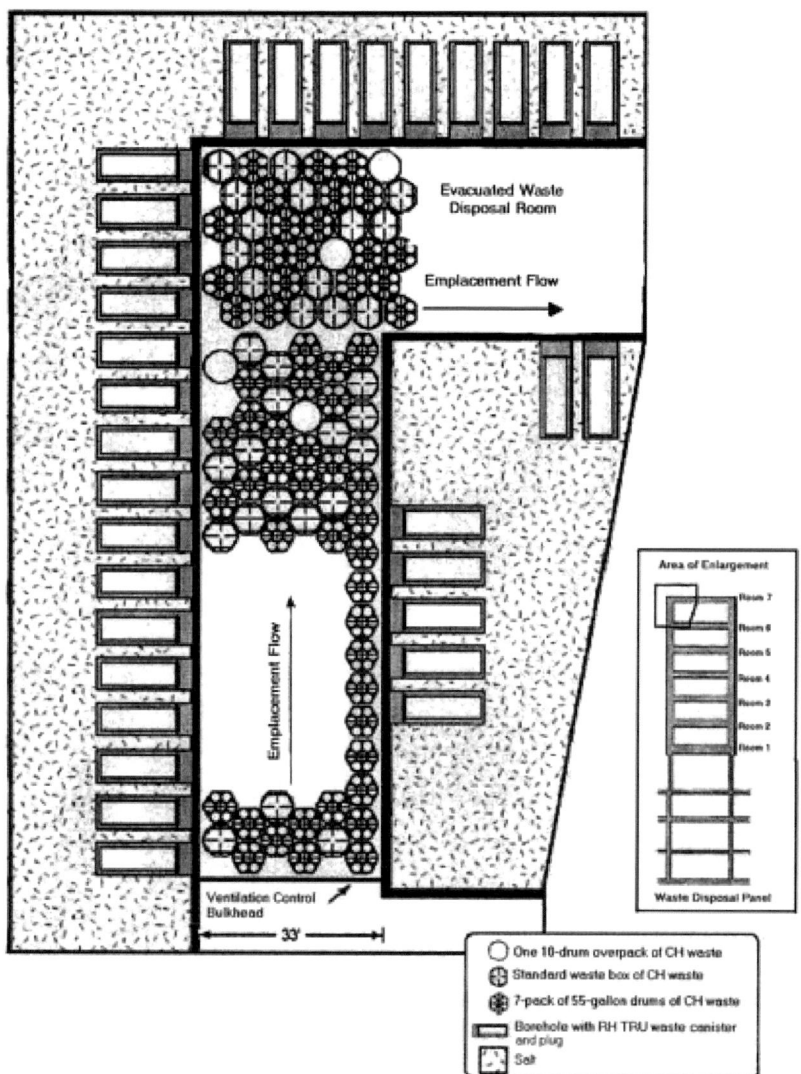

FIGURE 1.2 Plan view of a section of a typical underground disposal area showing CH-TRU and RH-TRU waste emplacement. SOURCE: National Transuranic Waste Management Program, Rev. 2, Figure 3.2.1.1–2 (DOE-CBFO, 2000; page 69).

Document 1 contains EPA regulatory requirements to ensure that the waste emplaced in WIPP falls within the envelope limits required to maintain compliance with radiological disposal standards. Document 2 contains regulatory requirements to ensure that the non-radiological hazardous characteristics of the waste emplaced in WIPP are compatible with RCRA requirements. Documents 1 and 2 also contain additional information, such as the WIPP's waste acceptance criteria, transportation requirements, description of surface storage and handling facilities, and waste certification process.

DOE's proposed characterization plan must address the characterization requirements in the EPA Certification and RCRA Permit in the context of RH-TRU waste. As shown in Chapter 2, this context is different from that of CH-TRU waste because of the different inventories and radiological properties.

EPA and NMED may accept, reject, or propose modifications to the characterization plan and eventually they may authorize RH-TRU waste disposal in WIPP. Once the characterization methods for RH-TRU waste are finalized, it is the committee's understanding that DOE will produce one document combining all regulatory requirements to help generator sites implement their characterization programs.

SIDEBAR 1.2 THE WASTE ISOLATION PILOT PLANT

The Waste Isolation Pilot Plant (WIPP), located near Carlsbad in New Mexico, is the nation's defense-related transuranic waste repository, as designated by Congress in the Land Withdrawal Act of 1992. This Act allows the disposal of 175,564 cubic meters of transuranic waste in WIPP, of which 7,080 cubic meters can be remote-handled transuranic (RH-TRU) waste with a maximum radioactivity of 5.1 million curies. Other legal criteria for WIPP related to RH-TRU waste are discussed in Chapter 3.

The WIPP disposal area is located 660 meters below ground in a salt bed, called the Salado Formation (see Figure 1.1). Large salt beds such as this are found only in regions that lack significant flows of groundwater. This deep, relatively dry, underground environment greatly reduces the possibility of waste releases from the repository by natural processes. Moreover, after approximately 200 years, the mined salt will heal and encapsulate the waste, thereby permanently locking it deep beneath the surface (Knowles and Economy, 2000).

The underground waste disposal area in WIPP consists of eight panels, each containing seven rooms. Contact-handled transuranic (CH-TRU) waste drums and boxes are being stacked in three layers in the center of each room. Remote-handled TRU waste is currently not authorized in WIPP. Should the U.S. Environmental Protection Agency (EPA) and the New Mexico Environment Department (NMED) allow disposal of RH-TRU waste in WIPP, it will be emplaced inside horizontal boreholes drilled into the walls of each room. Figure 1.2 shows the CH-TRU and RH-TRU waste emplacement configurations. According to WIPP's design, the diameter of a borehole for RH-TRU waste containers is 76.2 cm and the length is 487.68 cm (DOE-CAO, 1995). A shield plug will cap each RH-TRU waste borehole to provide workers with the necessary protection against radiation.

The WIPP has been under study since the mid-1970s and under construction since January 1981. The facility received the first CH-TRU waste shipment in March 1999 and the first mixed CH-TRU waste shipment in September 2000. To date (July 2002), more than 900 shipments of CH-TRU waste from 26 generator sites have been sent to WIPP.

After three years of operation, the first panel has been almost completely filled with CH-TRU waste. Because CH-TRU waste now blocks the access to the rooms walls, RH-TRU waste cannot be disposed in this first panel. Emplacement of CH-TRU waste in the second panel is currently scheduled to begin in December 2002.

INTRODUCTION

1.1 COMMITTEE'S TASK AND ITS BOUNDARIES

At the request of DOE-Carlsbad Field Office, the National Research Council appointed the Committee on the Characterization of Remote-Handled Transuranic Waste for the Waste Isolation Pilot Plant, referred to as "the committee" in the rest of this report. The committee roster and biographical sketches can be found in Appendix A. The committee's statement of task is reproduced in Sidebar P.1 of the Preface. The committee followed the examples of criteria listed in the statement of task to assess DOE's proposed characterization plan.

The committee was asked to provide recommendations, as necessary, for improving the characterization plan's technical soundness, protection of worker safety and health, and compliance with regulatory requirements. The difficulty of this latter part of the task lies in performing a review of a characterization plan for RH-TRU waste that must show compliance with existing characterization requirements. These requirements exclude RH-TRU waste. Moreover, the approach currently used to characterize CH-TRU waste is different than the approach DOE proposed for RH-TRU waste. It is important to emphasize the distinction between regulatory requirements and the approach to meet such requirements. While the regulatory requirements for the characterization of TRU waste destined for WIPP, established in the EPA Certification and RCRA Permit, apply to both CH- and RH-TRU waste, the waste characterization plans may propose different approaches to address these requirements. The committee evaluated, from a technical point of view, the approach DOE is proposing to characterize RH-TRU waste and how DOE proposes to address regulatory requirements. The committee was not asked to comment on these requirements nor was it asked to determine if the plan complies with the regulatory requirements. The latter is obviously a policy decision belonging to the regulatory agencies.

During this study, the committee reviewed two drafts of Documents 1 and 2, dated July 2001 (DOE-CBFO, 2001a, 2001b) and March 2002 (DOE-CBFO, 2002a, 2002b). The committee provided initial findings and recommendations in an interim report (see excerpt in Appendix C).

The committee did not verify data provided by DOE on RH-TRU waste inventories. Acknowledging the past fluctuations of RH-TRU waste inventory information, the committee assumes that DOE presented the most up-to-date information available. The committee was not asked to identify other potential sources of RH-TRU waste or discuss the origin of RH-TRU waste streams at generator sites.

Also, the committee did not assess the validity of supplementary information supporting the characterization plan, such as the Sandia Inventory Impact Assessment Reports in Document 1 (DOE-CBFO, 2002a; Attachment 2) and Document 2 (DOE-CBFO, 2002b; Supplement 2). The committee did, however, assess how DOE interpreted and used the results of these reports in its characterization plan.

This document is not meant to be a comprehensive review of the entire RH-TRU waste program, which encompasses RH-TRU waste transportation, storage, waste certification, operational safety issues, occupational health and safety regulations, waste generator states' regulations, waste acceptance criteria, as well as DOE orders. The National Research Council is undertaking a more general study on the characterization of TRU waste (including relevant transportation requirements) and its impact on DOE's National Transuranic Waste Management Program.

Documents 1 and 2 were submitted to the regulatory agencies on June 28, 2002. The committee did not review the submittal version of these documents; therefore, findings and recommendations in this report apply to the characterization plan as presented in March 2002.

1.2 ORGANIZATION OF THE REPORT

Following this introduction, Chapter 2 contains updated information on life cycle, status, and inventories of RH-TRU waste at selected generator sites. Chapter 3 describes the regulatory context for the disposal of RH-TRU waste. DOE's proposed characterization plan for RH-TRU waste is described in Chapter 4. Chapter 5 provides the committee's assessment of DOE's proposed characterization plan. In December 2001, the committee released an interim report with initial findings and recommendations relevant to the July 2001 draft. These findings and recommendations can be found in Appendix C. DOE's response to the committee's interim report is reproduced in Appendix D.

2

Remote-Handled Transuranic Waste

This chapter provides the context of RH-TRU waste characterization. The information presented on isotopic composition, life cycle, generator sites, and inventories for RH-TRU waste is provided on the basis of the knowledge gathered during committee meetings and has not been verified. DOE's inventories have been updated since the committee's interim report (released in December 2001).

2.1 TECHNICAL CONTEXT OF REMOTE-HANDLED TRANSURANIC WASTE CHARACTERIZATION

The radiation-related health hazard associated with TRU waste is due primarily to alpha and gamma radiation. Alpha radiation cannot penetrate human skin but poses a potential health hazard if particles containing alpha-emitting radionuclides are inhaled or ingested. Contact-handled TRU waste typically emits relatively little gamma radiation; therefore, when properly packaged, it can be handled directly by workers.

Remote handled-TRU waste also contains activation and fission products (half-lives are indicated in parenthesis), such as cobalt-60 (5.3 years), plutonium-241 (14.4 years), strontium-90[1] (29 years), cesium-137 (30 years), and their progenies. Some of these products emit gamma radiation, which can penetrate human skin and even the walls of waste containers. Therefore, RH-TRU waste requires heavy shielding and remote-handling equipment. Gamma rays represent the main radiological health hazard to workers during normal RH-TRU waste handling operations.[2] Although alpha-radiation exposure has a greater health risk per unit of energy deposited, the likelihood that workers will be exposed to alpha radiation while handling RH-TRU waste is lower compared to the likelihood for exposure to gamma radiation, because alpha radiation is stopped by the container's walls.

In summary, RH-TRU waste presents a long-term (i.e., 10,000 years) radiological hazard associated with TRU radionuclides (related to the presence of long-lived alpha-emitting isotopes) and a short-term (less than 300 years) radiological hazard associated with short-lived gamma-emitting radionuclides (related to the presence of fission and activation products). Today, most of the radioactivity in DOE's inventory of RH-TRU waste is due to the short-lived non-TRU radionuclides listed above. Nearly all of this activity will decay in 300 years, corresponding to approximately 10 half-lives of the

---

[1] Strontium-90 and plutonium-241 also emit beta particles (contributing to approximately 35 percent of the curie inventory for RH-TRU waste) but their energy is not high enough to penetrate the walls of RH-TRU waste containers.

[2] Normal waste handling operation implies that the RH-TRU waste container is not breached and workers are not directly exposed to alpha-emitting particles.

gamma-emitting radionuclides. Therefore, the long-term activity of RH-TRU waste is due to the remaining long-lived alpha-emitting (i.e., TRU) radionuclides.

2.2 LIFE CYCLE OF REMOTE-HANDLED TRANSURANIC WASTE

The life cycle of RH-TRU waste, from its generation to its designated final disposal at WIPP, generally consists of the following:

1. waste generation or recovery from its current storage location;
2. processing;
3. characterization;
4. packaging or repackaging;[3]
5. storage on site or off site prior to shipment to WIPP;
6. road transportation to WIPP;
7. receipt, handling, surface interim storage; and
8. underground emplacement at WIPP.

Remote-handled waste is generated during deactivation and decommissioning activities of DOE sites or it is already stored at generator sites and must be recovered for shipment to WIPP. Recovery activities can be complex if containers are difficult to access, if their integrity has been compromised with time, or if storage records are not available. For instance, part of the RH-TRU waste stored at Oak Ridge National Laboratory consists of a sludge that must be processed before shipment to WIPP (see Appendix E).

Waste processing activities consist of converting waste to a form suitable for disposal in WIPP. For instance, wet sludge must be dried because liquids are considered prohibited items (if more than 1 percent by waste volume) in WIPP's RCRA Permit. Waste characterization is necessary prior to waste shipment to WIPP. Waste characterization may be performed at the time of packaging or repackaging of the waste. Packaging is necessary for waste that has not yet been generated. Repackaging is necessary for stored waste that is packaged in a container that does not meet transportation requirements; or for waste that needs processing to remove hazardous material (see Section 2.4). Solid RH-TRU waste will be repackaged in shielded facilities, such as hot cells.[4] During this step, waste can be inspected by visual examination. As the waste is removed from an old container and sorted into new containers, a video camera records the operations on tape while an operator describes the various items. This inspection method requires skilled operators since it is done entirely by remote methods (see Finding 2D in Chapter 5). Volume-reducing operations, such as mechanical waste compacting, are also performed during the repackaging process.

Once a waste stream[5] is characterized and packaged, it must be certified by EPA and NMED and stored on site prior to shipment to WIPP. Selected waste streams may

---

[3] For most of the RH-TRU waste inventory, characterization will be performed at the time of packaging or repackaging of waste (see Chapters 4 and 5).

[4] A hot cell is a large chamber for handling highly radioactive materials. It is usually equipped with thick walls with shielding and viewing windows, remote-operated overhead cranes, closed-circuit televisions, and a variety of specialized tools and measuring devices. Throughout DOE complex other shielded facilities may be used with the same purpose as hot cells.

[5] A waste stream is defined as waste material generated from a single process or activity or as waste in multiple containers having similar physical, chemical, or radiological characteristics.

not need certification if they are shipped to other major DOE sites for interim storage. Prior to shipment to WIPP, these interim storage sites will ensure certification of all their waste, including that received from other sites.

Finally, RH-TRU waste must be loaded into transportation casks, transported by road[6] to WIPP, received, temporarily stored at the surface, and finally emplaced underground.

2.3 GENERATION OF REMOTE-HANDLED TRANSURANIC WASTE

According to DOE's latest inventory, there are approximately 3,800 cubic meters of RH-TRU waste to be removed from DOE sites across the nation. Figure 2.1 shows the geographic locations of the major RH-TRU waste generator sites. These sites are the following:

1. Hanford Site, Washington;
2. Idaho National Engineering and Environment Laboratory;
3. Los Alamos National Laboratory, New Mexico; and
4. Oak Ridge National Laboratory, Tennessee.

There are also nine smaller RH-TRU waste generator sites, not shown in Figure 2.1. Tables 2.1 and 2.2 show DOE's RH-TRU waste inventories in terms of volume and radioactivity. Additional information about the nature and history of RH-TRU waste at selected generator sites is provided in Appendix E.

Tables 2.1 and 2.2 show the substantial variability among generator sites concerning waste volumes and radioactivity contents. These tables show that 60 percent of the stored volume of RH-TRU waste and about 89 percent of the total curie activity is located at Oak Ridge National Laboratory. About two-thirds of the Oak Ridge RH-TRU waste consists of wet sludge, stored in Melton Valley storage tanks, and the rest is debris waste. Debris waste consists of debris from hot cells and glove boxes packaged in shielded concrete casks. This wet sludge and debris waste contains significant sources of neutrons, due to the spontaneous fission of californium-252 and from (alpha, n) reactions (see Appendix E). These reactions involve the capture of highly energetic alpha particles, such as those emitted from curium-244, by nuclei of low-atomic-number elements, such as oxygen and fluorine, and the subsequent emission of neutrons. The neutron emission is primarily a handling, processing, and particularly a characterization issue in the near term for this site. The waste containers that will be used to ship RH-TRU waste from Oak Ridge to WIPP are specially designed to shield neutrons.

To meet the milestones set in the Federal Facility Compliance Act between DOE and the state of Tennessee, the Oak Ridge National Laboratory (through its contractor Foster Wheeler) is building a treatment facility to process, characterize, and package its RH-TRU waste (EPA, 1992). The current plan is to characterize RH-TRU waste following a plan similar to that for CH-TRU waste characterization. Homogeneous samples of wet sludge would be characterized by radiochemical assays performed in a laboratory on site. The sludge would then be dewatered and dried in preparation for shipment. Debris would be characterized by visual and non-destructive examination[7] inside a hot cell

---

[6]The U.S. Nuclear Regulatory Commission certifies the design of transportation casks. Currently, these containers are designed for road transportation only.

[7]The definition of non-destructive examination can be found in the glossary, Appendix I.

during repackaging. Debris waste would undergo volume-reduction operations during packaging. Oak Ridge records show that there is considerable historical information available on debris waste (ORNL, 1989).

The processing of RH-TRU waste sludge had been scheduled to begin in December 2002 with shipments to WIPP beginning in January 2003. Debris RH-TRU waste would have been processed and shipped to WIPP between 2004 and 2008. Because of the recent submission of DOE's characterization plan to EPA and NMED (June 28, 2002), the characterization plan will likely not be finalized in time to meet this schedule. Oak Ridge is now organizing waste processing and on-site storage as it waits for authorization to ship RH-TRU waste to WIPP (Forrester et al., 2002).

While most of the stored RH-TRU waste volume is located at Oak Ridge National Laboratory, most of the RH-TRU waste to be emplaced in WIPP has yet to be generated. This waste will come from the Hanford Site cleanup activities. The RH-TRU waste at this site consists mostly of sludge at the bottom of the fuel pools, or equipment inside waste tanks (such as pumps, mixers, and pipelines) or is buried in caissons and drums. Because this waste has yet to be generated and packaged, the Hanford Site has less detailed knowledge of its RH-TRU waste inventory than the other major generator sites.

At the Idaho National Engineering and Environmental Laboratory, the stored RH-TRU waste consists mostly of solids generated during the destructive examination of irradiated experimental fuel pins in a hot cell. The RH-TRU waste stored at this site was generated at Argonne National Laboratory-East from defense-related experiments on nuclear fuel. The destructive examination and testing operations of spent fuel pins involved cutting, grinding, and polishing for subsequent examination.

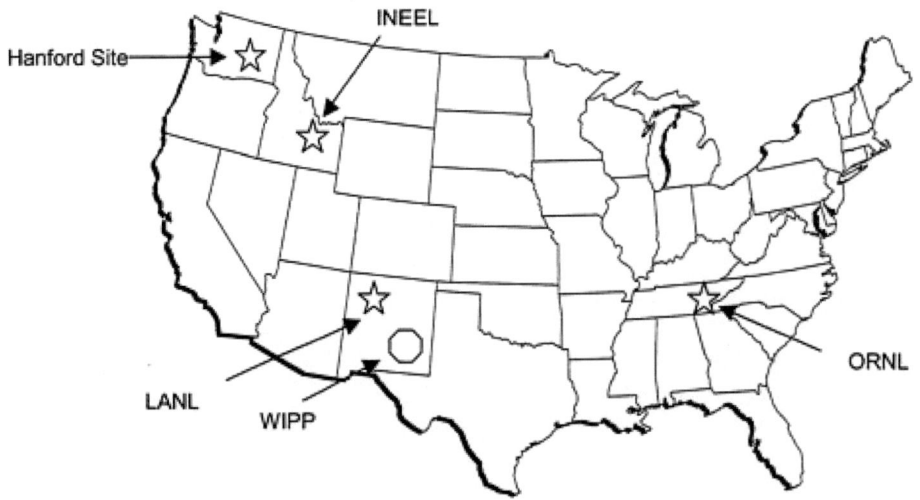

FIGURE 2.1 Geographic locations of the Waste Isolation Pilot Plant (WIPP) and the four major RH-TRU waste sites: Hanford Site, Idaho National Engineering and Environmental Laboratory (INEEL), Oak Ridge National Laboratory (ORNL), and Los Alamos National Laboratory (LANL). There are nine other smaller RH-TRU waste generator sites and several other sites throughout the nation that store CH-TRU waste. SOURCE: DOE (DOE-CBFO, 2001c).

TABLE 2.1 DOE's Inventory of Stored, Projected, Total, and Planned Volumes of RH-TRU Waste

Site Name	RH TRU Waste Volume (cubic meters)				Estimated Range of Concentration of CPR[5] (average kg per cubic meter)	Reported Estimates of Dose Rates[6] (rem per hour)
	Stored[1]	Projected[2]	Total[3]	Planned Disposal[4]		
Hanford Site	207	944	1,151	1,048	0–477 (24)	0.2–1,000
Idaho National Engineering and Environmental Laboratory	84	52	136	279	0–439(41)	0.2–100
Los Alamos National Laboratory	98	24	122	122	0–980 (95)	1–100
Oak Ridge National Laboratory	1,308	534	1,841	453	0–902 (74)	0.2–1,000
Total Small-Quantity Sites	1,697	1,554	3,250	1,902		
Argonne National Laboratory-East[8]	2	8	10	10	NA	0.2–10
Argonne National Laboratory-West	1.1	5	6.1		0–1,350(110)	1–100
Battelle Columbus Laboratories[8]	0	20.8	20.8	20.8	0–1,430(117)	0.2–150
Bettis Atomic Power Laboratory	2	0	2	2	0–1,430(117)	1–100
Energy Technology Engineering Center[8]	8	0	8.7	5.5	0–74 (5.2)	0.2–1
General Electric-Vallecitos Nuclear Center	11.8	0	11.8	11.8	NA	1–100
Knolls Atomic Power Laboratory	3.1	6.8	9.9	10.5	NA	1–100
Sandia National Laboratories[8]	1.5	22	23.5		NA	NA
West Valley Demonstration Project[9]	470.5	8.4	478.9		NA	NA
Total Waste Volumes	2,197.7	1,625	3,821.7	1,962.6		

NOTE: The committee did not verify the information in this table.
[1]Stored=waste awaiting treatment or disposal capability, in such a manner as to constitute disposal of the waste.
[2]Projected=the part of the inventory that has not been generated but is currently estimated to be generated at some time in the future.
[3]Total=sum of the stored plus projected volumes.
[4]Planned Disposal=volume expected to be disposed at WIPP. The quantities reflect any volumetric change that would occur during waste processing.
[5]Dose Rate Ranges=the dose rates estimated for as-packaged containers. Dose rates are measured at the container's surface.
[6]CPR=cellulosics, plastics, and rubber. A zero in the average CPR concentration indicates a container including other materials, such as metals or sludge.
[7]It is estimated that about 896 cubic meters of waste will be shipped from Idaho Nuclear Technology and Engineering Center for disposal at WIPP. This waste is expected to be categorized as "Waste Incidental to Reprocessing" and has not yet been approved for disposal at WIPP. Therefore, it is not included in the table.
[8]Current plans are to send the RH-TRU waste at this site to an interim site. Los Alamos National Laboratory's inventory includes Sandia's RH-TRU waste disposal volume. Argonne National Laboratory-East, Battelle Columbus Laboratories, and Energy Technology and Engineering Center are also expected to go to interim sites, but the specific interim sites have not yet been determined. Therefore, the waste volume is included here for completeness.
[9]The West Valley Demonstration Project is not a DOE site, even though it stores some RH-TRU waste. This waste has not yet been determined to be a defenserelated waste. It is believed to be commercially derived and therefore not disposable at WIPP in accordance with the Land Withdrawal Act. However, for completeness, this waste is included in the RH-TRU waste inventory.
SOURCE: Document 1, Supplemental Information, Attachment 3, Table 1 (DOE-CBFO, 2002a; page 5).

TABLE 2.2 DOE's Summary Activity Estimates for the RH-TRU Waste Inventory

Site Name	Estimated Stored Activity (total curies)
Hanford Site	36,000
Idaho National Engineering and Environmental Laboratory	6,360
Los Alamos National Laboratory	10,700
Oak Ridge National Laboratory	587,000
Small-Quantity Sites	
Argonne National Laboratory-East	NR
Argonne National Laboratory-West	NR
Battelle Columbus Laboratories	5,800
Bettis Atomic Power Laboratory	16,300
Energy Technology Engineering Center	8
General Electric-Vallecitos Nuclear Center	NR
Knolls Atomic Power Laboratory	118
Sandia National Laboratories	NR
West Valley Demonstration Project	NR
Total Waste Activity	662,286

NOTE: These estimates are for waste currently in storage. They do not take into account "to be generated" RH-TRU waste. The current estimate for stored and future RH-TRU waste is about 1 million curies. NR=Not Reported. NR typically denotes that the site did not report the information for a variety of reasons (e.g., the data are not readily available, the pedigree of the data may be questionable, recent characterization has not been performed, or the site did not respond). The committee did not verify the information in this table.
SOURCE: Document 1, Supplemental Information, Attachment 3, Table 2 (DOE-CBFO, 2002a; page 6).

A fine fuel participate adhered to the tools required for these operations and contaminated them with TRU elements. These tools (e.g., grinding and cutting wheels, glassware, light bulbs, rags) have been sent to the Idaho National Engineering and Environmental Laboratory. A formal determination of the defense origin for this waste stream will be submitted for approval to DOE prior to disposal at WIPP (Bhatt, 2001). The fuel pin segments, integral fuel pins, and the fuel dust are still stored at Argonne National Laboratory-East, as high level waste or spent nuclear fuel.

Most of the RH-TRU waste stored at Los Alamos National Laboratory was characterized and packaged between 1989 and 1994. The waste was characterized using a similar approach to that approved for CH-TRU waste characterization. Visual examination, non-destructive assay methods, radiography, and radiochemical analyses were used. Characterization operations were performed in a hot cell, except for small quantities of waste analyzed in a laboratory for radiochemical composition.

Among the nine smaller generator sites, the Battelle Columbus Laboratories in Ohio is the only site that is actively characterizing RH-TRU waste. This site is the first originally scheduled to ship RH-TRU waste to WIPP (January 2003). The agreement between Battelle Columbus Laboratories and the state of Ohio requires RH-TRU waste removal from the site by 2002 to meet a 2006 decontamination and decommissioning-completion milestone (Biedscheid et al., 2002).

Battelle Columbus Laboratories began RH-TRU waste characterization activities in October 2001 following a characterization program similar to that used for CH-TRU waste. This characterization program relies on visual examination, performed during waste repackaging, and previous knowledge of the waste, also called "acceptable knowledge" (AK). Acceptable knowledge is discussed in Sidebar 2.1. The repackaging phase, which occurs in a shielded facility, is necessary because RH-TRU waste containers at this site do not meet transportation requirements. Visual examination is used to estimate physical waste parameters, including weight percentages of metals, cellulosics, plastics, and rubber in the waste, and to determine the absence of prohibited items, including free liquids (see Chapter 3). Acceptable knowledge combined with computer modeling is used to estimate radiological waste parameters, including total activity on a waste container basis (see Appendix G). Acceptable knowledge combined with radiochemical analysis is also used on a fraction of RH-TRU waste for confirmation purposes. For a detailed description of the Battelle Columbus Laboratories' characterization program see Biedscheid et al. (2002).

This site is now seeking interim storage for its RH-TRU waste at the Hanford Site until shipments to WIPP are authorized. Work on a memorandum of agreement among DOE offices in Ohio, Carlsbad, and Richland is in progress (DOE-BCL, 2002) as part of the Hanford Site cleanup acceleration program. One element of this program is a proposal to transform the Hanford Site into a TRU waste-processing center for small generator sites, including Battelle Columbus Laboratories.

2.3.1 Remote-Handled Transuranic Waste Volume Inventory

Table 2.1 shows that the projected total volume inventory of RH-TRU waste in WIPP is 3,821 cubic meters, of which 2,197 cubic meters are currently stored at DOE sites and 1,625 cubic meters are yet to be generated (see the column "Projected"). For reference, the total volume of RH-TRU waste would correspond to approximately 20,000 RH-TRU waste containers (assuming they are all 55-gallon drums). Further volume-reducing[8] operations (mainly at Oak Ridge National Laboratory) may decrease the waste volume to be emplaced in WIPP to 1,963 cubic meters, as shown in the "Planned Disposal" column. According to these data, RH-TRU waste volume inventory represents between 1 and 2 percent of the total TRU (CH-TRU plus RH-TRU) volume allowed in the WIPP facility (175,564 cubic meters). The volume of RH-TRU waste to be emplaced in WIPP is also well below the regulatory limit of 7,080 cubic meters set by the Land Withdrawal Act.

There have been substantial variations in the estimated volumes of RH-TRU waste in DOE weapons complex, as also observed by the New Mexico Environmental Evaluation Group (EEG, 1994). These variations arise mostly from changes in the RH-TRU waste management or treatment plans. For example, DOE's earlier data for the Savannah River Site, in South Carolina, originally indicated a large RH-TRU waste inventory, reflecting disposal of materials stored in the separations canyons as TRU waste. However, the Savannah River Site now plans to send this material to the high-level waste tanks and process it (i.e., transform it into a glass waste form) with other high-level wastes for eventual disposal in a federal high-level waste repository. DOE also indicated that other potential sources of RH-TRU waste at the Idaho National Engineering and Environmental Laboratory are currently being reviewed (DOE-CBFO, 2002b; Supplement 1; page 2). Other potential changes in the RH-TRU waste inventory are noted in Table 2.1. The uncertainties in the estimated volumes can also be significant for the sites that have yet to generate their RH-TRU waste, such as Hanford.

---

[8]Packaging of Idaho National Engineering and Environmental Laboratory RH-TRU waste will actually increase its total volume, as indicated in Table 2.1.

SIDEBAR 2.1 ACCEPTABLE KNOWLEDGE

The Environmental Protection Agency's (EPA's) interpretation of acceptable knowledge (AK) is associated with historical information about the waste. The following is an excerpt of the guidance manual for waste analysis at facilities that generate, treat, store, and dispose of hazardous wastes (EPA, 1994):

"Wherever feasible, the preferred method to meet the waste analysis requirements is to conduct sampling and laboratory analysis because it is more accurate and defensible than other options. […] However, generators and TSDFs [treatment, storage, or disposal facilities] also can meet waste analysis requirements by applying AK. Acceptable knowledge can be used to meet all or part of the waste analysis requirements.

Acceptable knowledge can be broadly defined to include:

- 'Process knowledge,' whereby detailed information on the wastes is obtained from existing published or documented waste analysis data or studies conducted on hazardous wastes generated by processes similar to that which generated the waste. […] Therefore, with many listed wastes the application of AK is appropriate because the physical/chemical makeup of the waste is generally well known and consistent from facility to facility.
- Waste analysis data obtained from facilities which send wastes off site for treatment, storage, or disposal (e.g., generators).
- The facility's records of analysis performed before the effective date of RCRA regulations. While seemingly attractive because of the potential savings associated with using existing information (such as published data), the facility must ensure that this information is current and accurate (pages Introduction-11 and Introduction-12). […]

Generators and treatment, storage, or disposal facilities may use AK alone or in conjunction with sampling and laboratory analysis. […] However, there are situations where it may be appropriate to apply AK, including:

- Hazardous constituents in wastes from specific processes are well documented […].
- Wastes are discarded unused commercial chemical products, reagents or chemicals of known physical, and chemical constituents. […]
- Health and safety risks to personnel would not justify sampling and analysis (e.g., radioactive mixed waste).
- Physical nature of the waste does not lend itself to taking a laboratory sample" (pages Introduction-13 and Introduction-14).

For remote-handled waste characterization, the Department of Energy adopts a broader definition of AK, based on historical and new information, as explained in Chapter 4.

2.3.2 Remote-Handled Transuranic Waste Radioactivity Inventory

Table 2.2 (see above) shows that the activity of RH-TRU waste currently in storage is approximately 660,000 curies. DOE calculated the total activity of the RH-TRU (stored plus projected) waste inventory to be approximately 1 million curies (DOE-CAO, 1996a; Appendix BIR). According to this estimate, the amount of RH-TRU curies to be emplaced in WIPP is about 20 percent of the total amount of curies from RH-TRU waste allowed by the Land Withdrawal Act (5.1 million curies). For comparison, the total activity to be emplaced in WIPP from CH- and RH-TRU waste is estimated to be 7.4 million curies, of which 6.4 million come from CH-TRU waste (DOE-CAO, 1996c; Appendix BIR).

Table 2.3 shows a comparison between the principal TRU and non-TRU radionuclides in CH and RH-TRU waste to be emplaced in WIPP. In 1995, the total activity of RH-TRU waste represented approximately 14 percent of the total activity of the TRU inventory (from CH-TRU and RH-TRU waste). However, the activity due to TRU radionuclides in RH-TRU waste corresponds only to 0.5 percent of the total TRU activity expected in WIPP. Table 2.3 also shows a comparison of the isotopic characteristics of CH- and RH-TRU waste expected inventories in WIPP. The total curie inventory in RH-TRU waste represents roughly 16 percent of the total curie inventory in CH-TRU waste but the number of TRU curies in the latter is two orders of magnitude higher than in RH-TRU waste. Only TRU radionuclides (i.e., radionuclides with a half-life greater than 20 years) have an impact on the long-term[9] performance of WIPP since non-TRU radio

TABLE 2.3 Principal TRU and Non-TRU Radionuclides in WIPP Disposal Inventory

Radionuclide (half-life)	CH-TRU Waste (total curies)	RH-TRU Waste (total curies)	Ratio RH/CH
Plutonium-238 (87.7 years)	2,610,000	1,000	0.0004
Plutonium-239 (24,100 years)	785,000	10,000	0.013
Plutonium-240 (6,560 years)	210,000	5,000	0.024
Plutonium-241 (14.4 years)	2,310,000	142,000	0.061
Americium-241 (433 years)	442,000	6,000	0.014
Cesium-137 (30.2 years)	8,000	216,000	27
Barium-137m (2.6 minutes)	8,000	204,000	25.5
Cobalt-60 (5.3 years)	0	10,000	N/A
Strontium-90 (29.1 years)	7,000	209,000	30
Yttrium-90 (64 hours)	7,000	209,000	30
Total TRU curies (>20 years)	4,048,000	22,000	0.005
Total curies (TRU and non-TRU)	6,390,000	1,012,200	0.158

NOTE: Activities for RH- and CH-TRU waste were estimated in 1995, the year of inventory compilation. Activities decrease very rapidly for RH-TRU waste because of the short half-lives of its radionuclides. Total curies calculated assuming a volume of 7,080 cubic meters for RH-TRU waste and 168,500 cubic meters of CH-TRU waste. TRU radionuclides, by definition, have a half-life greater than 20 years.
SOURCE: Appendix BIR, Table 3–1 in TWBIR Rev 3. Supplemental Disposal Inventory Information, June (DOE, 1996c).

---

[9]Long-term refers to the 10,000 years of regulatory compliance with radionuclide release limits established by the EPA in Title 40 Code of Federal Regulations Part 191 (see Chapter 3).

nuclides will decay away in approximately 300 years.[10]

Therefore, RH-TRU waste has a negligible impact on the long-term performance of WIPP compared to CH-TRU waste because of the small volume to be emplaced in WIPP. However, RH-TRU waste presents a significant short-term potential for radiation exposure to workers during the waste characterization, handling, and emplacement period because of the short-lived gamma-emitting radionuclides.

2.4 PACKAGING STATUS OF REMOTE-HANDLED TRANSURANIC WASTE

A different way of sorting the RH-TRU waste inventory (other than the four categories in Table 2.1) is to consider its packaging status. As previously mentioned, before TRU waste is shipped to WIPP, it must be packaged in a suitable form for transport and underground emplacement.

In the March 2002 draft of DOE's characterization plan, the RH-TRU waste inventory is divided into four categories:

- "Packaged - Waste is packaged in a final form suitable for transport to and disposal at WIPP (i.e., the waste is either canisterized [Los Alamos National Laboratory RH-TRU waste] or in a package that may be transported for canisterization at another facility without repackaging).
- To Be Generated - Future waste generation, includes projected generation due to new activities and planned environmental restoration activities (i.e., retrieval from burial grounds, such as the 618 area at Hanford).
- To Be Repackaged - Waste is currently packaged in some form, but is not suitable for transport and disposal as packaged, possibly due to package size, package condition, or knowledge of contents. It will be repackaged into containers suitable for transport and disposal at WIPP. If a decision has already been made to package the waste for disposal at WIPP [i.e., the Oak Ridge National Laboratory RH-TRU waste debris], then this waste volume has been included in the To Be Packaged estimates.
- To Be Packaged - Waste is either not yet packaged (i.e., tank sludges) or the decision has been made to package the current form differently [i.e., RH-TRU waste debris waste at Oak Ridge National Laboratory]" (DOE-CBFO, 2002b; Supplement 1; page 2).

The current RH-TRU inventory of stored and projected waste is distributed among these four categories as shown in Figure 2.2. For comparison with Table 2.1, "Projected Waste" corresponds to the "To Be Generated" column. All other packaging situations correspond to the "Stored" column. The four packaging categories add up to the total number of cubic meters in Table 2.1. In its characterization plan, DOE also indicated that the inclusion of the Idaho Nuclear Technology and Engineering Center waste will further increase the "To Be Packaged" component by increasing the amount of waste in this category by approximately 900 cubic meters.

The committee's interim report discussed the RH-TRU waste inventory using DOE's initial distinction between "Retrievably Stored Waste" and "Newly Generated Waste"

---

[10]Short-lived radionuclides in RH-TRU waste are included in the calculation of the regulatory release limits. The scenario that would mostly affect release limits from RH-TRU waste is a drilling intrusion in WIPP occurring less than 300 years after closure. This scenario is taken into account in the performance assessment calculation (see Chapter 3, Section 3.2.1).

The committee's interim report discussed the RH-TRU waste inventory using DOE's initial distinction between "Retrievably Stored Waste" and "Newly Generated Waste" (NRC, 2001 b). Retrievably stored waste is any waste produced after 1970[11] but prior to implementation of an approved RH-TRU waste characterization plan (as of July 2002, no plan has been approved by EPA and NMED). Newly generated waste is waste produced after the development, approval, and implementation of a waste characterization plan and meets the requirements set forth by the regulatory agencies. In this report the committee adopts DOE's new classification system, as shown in Figure 2.2.

Earlier inventory information recorded that 80 percent of the waste was to be repackaged or generated, as cited in the committee's interim report (NRC, 2001 b). Updated information in the March 2002 draft indicated that 98 percent of the waste is to be generated or packaged (see Figure 2.2). However, in several other instances, Documents 1 and 2 report that the fraction of waste to be generated, packaged, or repackaged is 95 percent. In this report, the committee uses a conservative estimate of 95 (rather than 98) percent of waste to be packaged (for the first time or repackaged).

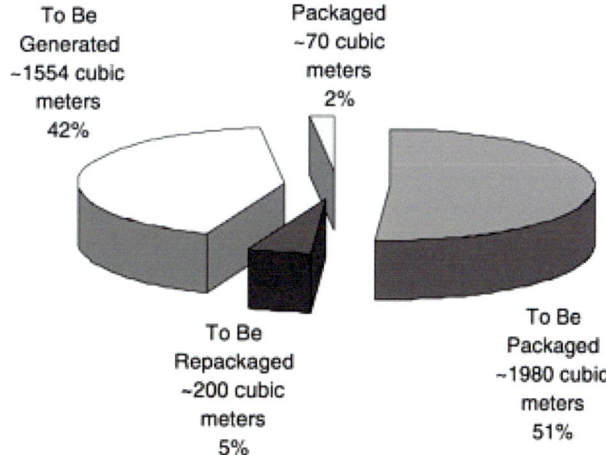

FIGURE 2.2 Packaging status of the RH-TRU inventory. SOURCE: DOE (DOE-CBFO, 2002a).

---

[11]In 1970, the defense operation division of the U.S. Atomic Energy Commission (predecessor to DOE) first identified TRU waste as a separate category of radioactive waste. That same year the Atomic Energy Commission determined that all TRU waste generated after 1970 must be segregated from low-level waste and placed in retrievable storage pending shipment to, and disposal in, an approved geological repository. Transuranic waste produced in support of the nuclear weapons program from the 1940s through 1970 was disposed of by shallow land burial and other techniques at a number of sites owned and operated by the federal government. This type of waste is referred to as "buried transuranic waste," and most of this waste is considered irretrievable. The characterization plan reviewed by the committee does not address buried TRU waste.

The RH-TRU packaged waste consists of waste from three sites: Hanford, Los Alamos National Laboratory, and Battelle Columbus Laboratories. Waste at Hanford and Battelle is stored in 55-gallon drums. At Los Alamos, waste is stored in 17 canisters, of which 16 were generated during program activities and 1 from hot cell cleanup after the work was completed. Though these canisters are packaged, they are not ready for shipment to WIPP because they have not been characterized in accordance with the current characterization requirements in the EPA Certification and RCRA Permit.[12] Nevertheless, thorough records of process knowledge, visual examination, and prohibited items determination (see Chapter 3, Section 3.3) were kept during the packaging of the 17 canisters stored at Los Alamos.

---

[12]For instance, when waste was packaged, the records do not demonstrate that the operator was qualified to interpret the videotape, and the headspace gases have not been collected.

3

Regulatory Context for the Disposal of Remote-Handled Transuranic Waste

The Land Withdrawal Act, Public Law 102–579, is the guiding legislation for WIPP (U.S. Congress, 1992). In this Act, Congress established the scope and legal criteria for the facility. The Land Withdrawal Act also assigned EPA regulatory authority over public exposure to radiation from the management and storage of TRU waste at WIPP during the operational period (nominally lasting until 2033) and over radioactivity releases into the environment up to 10,000 years after repository closure. The hazardous non-radioactive component of TRU waste is regulated by the NMED under RCRA (EPA, 1990). The NMED standards for a hazardous waste disposal facility (such as WIPP) seek to minimize releases[1] of non-radiological hazardous constituents during the operational phase and for a nominal 30-year period following repository closure.

The two main regulatory documents for WIPP are the EPA Certification and the RCRA Permit. To continue disposal operations, WIPP must show continued compliance with the EPA Certification and the RCRA Permit. One of the conditions contained in the EPA certification is that EPA and NMED must approve site-specific characterization measures and quality assurance programs before generator sites ship waste to WIPP. Therefore, the EPA Certification and RCRA Permit establish waste characterization requirements for all TRU waste (CH or RH) to be emplaced in WIPP.

The requirements for TRU waste characterization used by EPA and NMED originate from the following standards for owners and operators of hazardous waste treatment, storage, and disposal facilities:

> "(a)(1) Before an owner treats, stores, or disposes of any hazardous wastes, or nonhazardous wastes if applicable under §264.113(d), he must obtain a detailed chemical and physical analysis of a representative sample of the wastes. At a minimum, the analysis must contain all the information which must be known to treat, store, or dispose of the waste in accordance with this Part and Part 268 of this chapter" (40 CFR 264.13).

EPA indicates in the same document that characterization information can be obtained either by direct measurement or indirectly through AK (see Sidebar 2.1).

---

[1] With an exemption from Land Disposal Restriction regulations (40 CFR 268), Congress determined that WIPP waste does not need treatment prior to disposal. As a result, the NMED imposed groundwater and air monitoring requirements to ensure that any detectable release remains below specific limits that have been established to protect human health and the environment.

The regulatory requirements for the characterization of TRU waste in WIPP originate from the Land Withdrawal Act, the EPA Certification, and the RCRA Permit. These requirements are described below.

3.1 LAND WITHDRAWAL ACT REQUIREMENTS FOR TRANSURANIC WASTE IN WIPP

The Land Withdrawal Act establishes the five main legal requirements for TRU waste in WIPP, some of which apply specifically to RH-TRU waste. These requirements are the following:

- Nature of waste to be disposed of in WIPP: the WIPP can only receive transuranic waste generated by atomic energy defense activities.
- Total volume of transuranic waste: the WIPP is allowed to contain up to 175,564 cubic meters of transuranic waste.
- Volume of RH-TRU waste: the WIPP is allowed to contain up to 7,080 cubic meters of RH-TRU waste, which represents about 4 percent of the total volume of TRU waste allowed.
- Total activity of RH-TRU waste: RH-TRU waste received at WIPP shall not exceed 23 curies per liter maximum activity level (averaged over the volume of the container). The total curies of RH-TRU waste received at WIPP shall not exceed 5.1 million curies.[2]
- Surface dose rate: no TRU waste received at WIPP may have a surface dose rate exceeding 1,000 rems per hour. No more than 5 percent by volume of RH-TRU waste received at WIPP (354 cubic meters) may have surface dose rates exceeding 100 rems per hour.

Tolerable uncertainties are not specified in the Land Withdrawal Act for the above limits. Laws seldom specify accuracy or precision of limits; it is generally the duty of the implementer to propose tolerable decision errors, such as accuracy or uncertainty, derived from a tradeoff among relevant parameters, such as risk or costs.

3.2 CHARACTERIZATION REQUIREMENTS IN THE EPA CERTIFICATION

In 1998, EPA promulgated the criteria for the certification and recertification of the Waste Isolation Pilot Plant's compliance with the 40 CFR 191, subparts B and C (40 CFR 194). The EPA declared WIPP to be in compliance with its disposal regulations and granted the facility a certification of compliance. The purpose of the EPA characterization requirements is to ensure that WIPP remains in compliance with the EPA disposal regulations and limits set forth in the Land Withdrawal Act. Compliance is ensured if the waste inventory in WIPP stays within the waste envelope limits specified in WIPP's performance assessment. The performance assessment is a tool used to evaluate the safety performance of any nuclear waste facility, as explained below.

---

[2]Note that the total activity limits for RH-TRU waste do not distinguish between the activity due to the TRU nuclides and that due to other radionuclides.

3.2.1 WIPP's Performance Assessment

The performance assessment evaluates the ability of a repository to satisfy the release limits (listed in 40 CFR 191, Appendix A) on the release of radionuclides from a repository over a period of 10,000 years. The performance assessment organizes information relevant to long-term repository behavior by assessing the probabilities and consequences of major scenarios by which radionuclides can be released into the environment. Important scenarios include those due to human activities, whether deliberate or unintentional, that might compromise the integrity of the repository. The performance assessment process consists of:

1. compiling features, events, and processes that could affect the disposal system;
2. classifying events and processes to enhance consistency and completeness;
3. screening individual events and processes for likelihood of occurrence and consequences;
4. combining events and processes into specific scenarios;
5. screening scenarios to identify and eliminate those that have little or no effect on the performance assessment;
6. modeling the scenarios;
7. calculating the releases; and
8. comparing the calculated values with the allowable releases.

From the quantitative results of its performance assessment, DOE identified the ten radionuclides most important to the long-term performance of WIPP: americium-241, cesium-137, plutonium-238, plutonium-239, plutonium-240, plutonium-242, strontium-90, uranium-233, uranium-234, and uranium-238. Of these ten, strontium-90, uranium-233, and cesium-137 are important to RH- but not CH-TRU waste streams (due to the gamma and beta emissions).

A sensitivity analysis of the performance assessment also identified the non-radiological waste parameters most important to the performance of the repository (Helton et al., 1998). These parameters are the following:

- Amount of free water: The EPA Certification requirements include a limit on the total amount of water or brine present in the waste. This parameter is important because water is the only means to release radionuclides in the environment and because it controls several other factors, such as corrosion and gas generation.[3] The total limit for free water in WIPP is 1 percent by volume in the entire TRU waste inventory (175,564 cubic meters), which corresponds to 1,756 cubic meters. This limit derives from a transportation requirement forbidding shipment of hazardous waste containing more than 1 percent of free liquids by volume.[4] DOE used this requirement as one of the initial assumptions in the performance assessment calculation, which was part of the compliance certification application. This free water

---

[3] The main gases potentially generated in WIPP in the presence of brine are carbon dioxide, methane, and hydrogen. These gases are generated by microbial waste degradation inside WIPP. In addition, hydrogen can also be produced by radiolysis or by corrosion of metal containers. An increase in gas pressure inside the repository could affect room closure rates, fracture development, brine inflow, and the possibility of waste entrainment in gas during a drilling event (called spalling).

[4] The committee does not review transportation requirements in this report.

requirement is now integrated as part of the EPA Certification to ensure compliance with release limits.
- Amount of ferrous metals: The EPA Certification requirements include a minimum amount of ferrous metals (20 million kilograms) to ensure a reducing environment inside the WIPP repository. A reducing environment maintains radionuclides in a low oxidation state, which usually corresponds to a minimum solubility. Therefore, waste characterization must ensure that the amount of corrodible metals in the waste is above the minimum.
- Amount of non-ferrous metals: The EPA Certification requirements include a minimum amount (2,000 kilograms) of non-ferrous metals in WIPP. Non-ferrous metals form complexes with organic ligands present in waste (such as EDTA or citric acid[5]). This prevents actinides from binding with these organic ligands and becoming mobile. Therefore, waste characterization must ensure that the amount of non-ferrous metals in the waste is above the minimum.
- Amount of biodegradable cellulosics, plastic, and rubber: The EPA Certification requirements include a maximum amount (20 million kilograms) of cellulosics, plastic, and rubber in the waste to take into account the potential for gas generation from the decomposition of these organic materials. Therefore, waste characterization must ensure that the amount of cellulosics, plastic, and rubber in the waste is below the maximum.

Tolerable uncertainties have not been specified in the compliance certification application for the above limits. In its final certification decision, EPA writes:

"[T]he EPA confirmed the results of the performance assessment using the same upper and lower limiting values in the performance assessment verification test ("PAVT"). Those upper and lower limiting values apply to contact-handled, remote-handled, and to-be-generated waste from numerous generator sites. Thus, in today's action, EPA certifies that the WIPP will comply with the 40 CFR Part 191 containment requirements to the extent that emplaced waste falls within the waste envelope limits that were shown by the performance assessment, and confirmed by the PAVT, to be compliant with the 40 CFR Part 191 standards" (40 CFR 194).

In summary, the waste characterization requirements appearing in the EPA Certification are aimed to determine:

1. activity and radiological inventory of the waste;
2. surface dose rate;
3. amount of metals (ferrous and non-ferrous);
4. absence of free water; and
5. amount of cellulosics, plastic, and rubber.

Chapter 4 describes the approach DOE proposes to determine these 5 characterization parameters in RH-TRU waste.

---

[5]EDTA (ethylenediaminetetraacetic acid) and citric acid are common cleaning agents.

3.3 CHARACTERIZATION REQUIREMENTS IN THE RCRA PERMIT

The purpose of the NMED characterization requirements is to ensure that the non-radiological hazardous characteristics of the waste emplaced in WIPP are compatible with the RCRA requirements as codified in Part 20 of the New Mexico Administrative Code. The cornerstone of any RCRA permit is the waste analysis plan, which allows generators and disposal facilities to identify, treat, store, and properly dispose of all hazardous wastes. Waste analysis involves identifying or verifying the chemical and physical characteristics of a waste by performing a detailed chemical and physical analysis of a representative sample of the waste or, in certain cases, by applying AK (EPA, 1994).

The characterization plan in the RCRA Permit is aimed at minimizing the possibility of a fire, explosion, or any unplanned sudden or non-sudden release of TRU mixed waste or mixed waste constituents to air, soil, groundwater, or surface water which could threaten human health or the environment (see NMED, 1999a; Module II General Facility Conditions; page II–1).

Currently, the waste characterization requirements in the RCRA Permit are aimed at determining the following information:

1. physical form of waste (homogeneous solids, soil or gravel, debris); and
2. exclusion from waste of the following prohibited items[6]:

- Liquid waste.
- Non-radionuclide pyrophorics (addressed by the hazardous waste codes[7]).
- Hazardous wastes that do not contain TRU waste.
- Incompatible chemicals (addressed by the hazardous waste codes).
- Polychlorinated biphenyls greater than or equal to a concentration of 50 ppm.
- Explosives (addressed by the hazardous waste codes) and compressed gases.
- Ignitable, corrosive, and reactive wastes (addressed by the hazardous waste codes).
- Remote-handled transuranic mixed waste.

Prohibited items listed in the RCRA Permit are items that are incompatible (e.g., ignitable, reactive) or inappropriate for the WIPP facility. If combined, incompatible wastes are capable of spontaneous combustion, toxic gas generation, or explosions (EPA, 1994). The selection of waste parameters must include measures to ensure that the waste is appropriate for WIPP's scope and that inappropriate waste is identified and excluded prior to shipment.

The current RCRA Permit excludes RH-TRU mixed waste because:

"The Applicant [DOE] failed to submit an approvable waste analysis plan describing the procedures for obtaining a detailed chemical and physical analysis of RH [mixed] waste destined for disposal at WIPP. Moreover, there

---

[6]Additional prohibited items for WIPP are waste containers that do not have volatile organic compound concentrations reported for the headspace, waste containers that have not undergone radiographic or visual examination, and waste containers that do not have a certified waste stream profile form (see the glossary).

[7]The definition of hazardous waste codes can be found in the glossary, Appendix I.

are substantial questions regarding the applicability of CH waste characterization techniques and the Applicants' capability to characterize RH [mixed] waste" (NMED, 1999b).

To remove the RH-TRU waste prohibition from the RCRA Permit, NMED recommends that DOE build its proposed RH-TRU waste characterization plan upon the requirements in the existing permit and provide justification for any difference between the RH-TRU and the existing CH-TRU characterization plan (Zappe, 2001).

3.4 HISTORY OF CHARACTERIZATION REQUIREMENTS AT WIPP

When DOE first applied for the EPA certification of WIPP, it proposed the same characterization plan for RH-TRU and CH-TRU waste (DOE-CAO, 1996a). DOE, EPA, and NMED negotiated the characterization requirements for TRU waste during nearly two decades prior to the certification of WIPP. Records of the negotiation process show that DOE proposed to the regulators some waste characterization requirements that lacked a technical, safety, or legal basis (see Sidebar 3.1). These requirements eventually became part of the EPA Certification and RCRA Permit and are now addressed in the CH-TRU waste characterization plan.

Eventually, EPA and NMED declared that DOE did not provide an adequate waste characterization method for RH-TRU waste and prohibited this waste from being shipped to WIPP. Both agencies also stated that DOE did not adequately describe its full RH-TRU waste inventory. DOE is now submitting a new characterization plan for RH-TRU waste to address EPA and NMED characterization requirements. While the regulatory requirements to be addressed are the same for CH- and RH-TRU waste, DOE proposes a different approach to characterize RH-TRU waste than that used for CH-TRU waste, as described in the following chapter.

SIDEBAR 3.1 FINDINGS OF THE PREVIOUS NATIONAL RESEARCH COUNCIL'S COMMITTEE ON THE WASTE ISOLATION PILOT PLANT

A previous National Research Council committee on the Waste Isolation Pilot Plant (WIPP) found that some of the elements in the characterization plan for contact-handled transuranic (CH-TRU) waste are based on terms negotiated in the Resource Conservation and Recovery Act (RCRA) Permit rather than being performance-based characterization requirements (NRC, 2001a; Appendix A1). These requirements concern the sampling and analysis of homogeneous waste, headspace gas sampling and analysis, and visual examination procedures. For instance, the Department of Energy (DOE) proposed to conduct homogeneous waste and headspace gas sampling and analyses on the totality of CH-TRU waste containers to confirm acceptable knowledge (AK) information. Moreover, there appears to be no requirement in the law for the verification of radiography results by visual examination.

These requirements were proposed by DOE during the development of the characterization plan and "have no basis in law, the safe conduct of operations to emplace waste in WIPP, or long-term performance requirements" (NRC, 2001a; page 77). Therefore, the 1998 Committee recommended that DOE "eliminate self-imposed waste characterization requirements" from the CH-TRU waste characterization plan (NRC, 2001 a; page 78). DOE agreed with that committee on the fact that some waste characterization procedures are indeed not prescribed by safety or legal requirements and were introduced to facilitate the WIPP certification process (Kehrman, 2002; DOE-CAO, 1999a). A DOE review of the CH-TRU waste characterization procedures revealed that DOE developed self-imposed waste restrictions in the waste acceptance criteria and in the requirements for waste generating sites presented in the quality assurance program plan. The definitions of waste acceptance criteria (DOE-CAO, 1996b) and quality assurance program plan (DOE-CAO, 1998) can be found in the glossary, Appendix I.

4

Department of Energy's Proposed Characterization Plan

This chapter describes DOE's proposed characterization plan for RH-TRU waste. Because the plan must address EPA and NMED regulatory requirements, the characterization plan consists of two documents: a notification of proposed change to the EPA Certification (Document 1) and a Request for RCRA permit modification (Document 2). On June 28, 2002 DOE submitted Documents 1 and 2 to the respective agencies. This chapter addresses the March 2002 draft characterization plan and the main differences with the July 2001 draft.

It is the committee's understanding that DOE will submit to EPA and NMED, along with Documents 1 and 2, a sample of completed characterization documents for three RH-TRU generator sites (Energy Technology Engineering Center, Battelle Columbus Laboratories, and Oak Ridge National Laboratory). Once Documents 1 and 2 are submitted to the regulatory agencies, and characterization requirements for RH-TRU waste are finalized, DOE will produce one document combining all regulatory requirements to guide generator sites in the development of their characterization programs. Each site-specific RH-TRU waste characterization plan will undergo audits by DOE, EPA, and NMED before the beginning of shipments to WIPP. The new characterization requirements will be combined in a revised version of the waste acceptance criteria incorporating changes to allow RH-TRU waste in WIPP.

4.1 DOE'S CONSIDERATIONS IN COMPOSING THE CHARACTERIZATION PLAN

According to the information gathered, several factors played a role in DOE's approach to the characterization of RH-TRU waste, such as: DOE's mission to safely dispose of radioactive waste; WIPP's regulations; federal and state laws and agreements; and DOE's internal drivers to improve the National Transuranic Waste Management Program. The list of considerations, expanded below, is not in hierarchical order and may not be complete.

One of DOE's missions is to safely dispose of radioactive waste, including RH-TRU waste, generated during defense-related activities (DOE-EM, 2002a). To this end, DOE is seeking authorization to dispose of RH-TRU waste stored at 13 sites across the country in WIPP. The proposed characterization plan must meet the requirements in the EPA Certification and RCRA Permit (the latter issued by NMED). Besides strict regulatory requirements, DOE is also taking into account non-technical considerations to maintain an effective working relationship with the regulatory agencies, in particular NMED.

DOE and its RH-TRU waste generator sites must also meet sites' cleanup milestones, set forth by DOE's compliance agreements with EPA and generator states' regulators, such as Federal Facility Compliance Agreements and Tri-Party Agreements.

DOE is also required to comply with Title 10 of the Code of Federal Regulations Part 835, which instructs DOE to apply the ALARA principle to protect workers' health and safety in every DOE facility (10 CFR 835). ALARA (as low as is reasonably achievable), as defined by the Nuclear Regulatory Commission, requires that a reasonable effort be made to keep workers radiation exposures as far below the regulatory dose limit[1] as is practical. Moreover, DOE must ensure that all waste is managed in a manner that is protective of worker and public health and safety, and the environment, as specified in DOE's Order 435.1.

Timeliness and cost of cleanup are two internal DOE drivers. A "Top-to-Bottom" review report on DOE's Office of Environmental Management, which manages the cleanup of all defense-related radioactive waste, recommended adopting an accelerated risk-based cleanup strategy across the DOE complex. The following is an excerpt from this report:

> "National security will be improved through the consolidation of all special nuclear materials in modern safeguarded facilities and through accelerated disposal of transuranic (TRU) waste currently stored at sites around the country (page IV-1) [...] [A] limited amount of TRU waste is highly radioactive and must be handled remotely. Efforts to dispose of these materials in WIPP will be accelerated (page IV-2). [...] [T]here are thousands of TRU waste drums stored in above-ground EM facilities that require high-priority funding for safety and security. While most of these storage facilities are inexpensive to maintain, there are dozens of them across the complex, so the cumulative annual fixed cost is significant. Certification and disposal of TRU waste are major cost drivers for the EM program. Efforts to expedite shipments to the Waste Isolation Pilot Plant (WIPP) and streamline regulatory procedures would result in obvious cost savings" (page V-13) (DOE-EM, 2002b).

In its WIPP Disposal Phase Final Environmental Impact Statement, DOE estimated the life-cycle cost of the treatment facilities for RH-TRU waste to be $2.94 billion. This corresponds to 15 percent of the total cost of the $19 billion WIPP Project (DOE-CAO, 1997; Volume 5; pages 5–10). DOE also stated in the same reference that the life-cycle cost of the treatment facilities for all the CH- and RH-TRU waste was $11.8 Billion, or 62 percent of the total cost of the WIPP Project. Although these estimates may be outdated, they still demonstrate the magnitude of the characterization costs for TRU waste, in particular for RH-TRU waste.

4.2 DOE'S GOAL: A PERFORMANCE-BASED CHARACTERIZATION PLAN

Given the above considerations, DOE's stated goal is to adopt a "performance-based" approach to characterize RH-TRU waste. Although Documents 1 and 2 do not provide a definition of a "performance-based characterization plan," the committee found the following definition in EPA's Compliance Application Review Document No. 24:

> "DOE must provide waste inventory information for use in the performance assessment, including the radionuclide content of waste and the physical and

---

[1] The Nuclear Regulatory Commission's limits for radiation doses to workers are established in Title 10 CFR Part 20 (10 CFR 20.1003).

chemical components that may affect disposal system performance" (EPA, 1998; pages 21–1).

EPA defines a "waste characteristic" as "a property of the waste that has an impact on the containment of waste in the disposal system" (40 CFR 194.2). The committee interprets "performance-based characterization plan" as a plan requiring only that information needed to ensure repository integrity, protection of the environment, and health and safety of the public and workers while keeping in consideration tradeoffs between greater characterization accuracy and worker risks.

While the characterization requirements to be addressed are the same, this proposed performance-based approach is quite different from the approach used for CH-TRU waste characterization, as explained later in this chapter. The rationale for this different approach relies on the differences in volume and isotopic inventory of RH-TRU waste compared to CH-TRU waste (see Chapter 2).

To evaluate the impact of RH-TRU waste on the performance of WIPP, the Sandia National Laboratories performed two performance assessment calculations. In these calculations, the effects of the inventories of CH- and RH-TRU waste are decoupled and several bounding scenarios for RH-TRU waste parameters are used. Documents 1 and 2 include two impact assessments: one for the impact of the radiological parameters of RH-TRU waste and one for the non-radiological parameters listed in the EPA Certification.

The Sandia impact analyses showed that radioisotope, free liquids, cellulosics, plastics, rubber, and ferrous metal content in the RH-TRU waste inventory have a negligible impact on repository performance. This is due to the smaller volume and the different radiological composition of the RH-TRU waste inventory compared to the CH-TRU waste inventory projected in WIPP. These performance assessment calculations use the same mathematical models and input parameters used in the compliance certification application (see Chapter 3, Section 3.2.1).

The result of the calculations shows that the characterization of waste material parameter weights (e.g., ferrous metals, cellulosics, plastic, and rubber) and the amount of free liquid for RH-TRU waste can be assumed to be as high as the bounding values and still result in WIPP compliance. The bounding values for RH-TRU waste parameters are the same as those used for EPA Certification. The Sandia impact analyses also show that if all the total (both TRU and non-TRU) activity in RH-TRU waste were due to plutonium-239, there would be only a negligible performance impact on WIPP. Therefore, from a performance point of view, waste characterization activities to quantify the above waste parameters could be unnecessary.

Even though the impact of RH-TRU waste parameters on the performance of WIPP is negligible, DOE is proposing to collect all the waste information requested in the EPA Certification and RCRA Permit using AK as the primary method of compliance. When AK is not complete or not available, additional compliance methods can be applied to a representative portion of the waste. In contrast to the approach used for CH-TRU waste, this plan does not recommend making measurements for the sole purpose of confirming AK or other measurements unless such confirmation is needed to qualify the AK characterization information (DOE-CBFO, 2002a; page 22).

4.2.1. Acceptable Knowledge

The heart of DOE's proposed performance-based characterization plan is the use of AK to collect the necessary information about a waste stream. DOE plans to use two types of AK: one based on historical information of existing waste, which the committee terms "historical AK" and which is the most widespread interpretation of AK (see Sidebar 2.1), and one based on new information collected during waste generation or packaging.

Historical AK may include administrative, procurement, and quality control documentation associated with the generating process, or past sampling and analytical data. Elements of process knowledge may include information about the process used to generate the waste, material inputs to the process, and the time period during which the waste was generated. Acceptable knowledge is applied on a waste stream basis and may be supplemented with sampling and measurement programs or container-by-container measurements.

DOE also considers AK newly generated characterization information, as described in Document 1:

> "[F]or the RH-TRU waste characterization program, 95 [percent] of the waste will be packaged or repackaged during its preparation for shipment to the WIPP. Information collected under these packaging and repackaging programs is also considered to be AK information. This allows the same AK procedure and guidance to be used in assembling historic AK information or compiling new AK information" (DOE-CBFO, 2002a; Attachment B; page 23).

As shown above, DOE's rationale for proposing AK as the major compliance tool is based on the analysis of the RH-TRU waste inventory. According to the information gathered, 95 percent of RH-TRU waste will either be generated or packaged. This waste will be visually examined during repackaging operations and undergo physical and chemical characterization (for instance, through radiography or radiochemistry) during waste generation and packaging.

Based upon EPA's requirements set forth in 40 CFR 194.22(b), existing information collected before an approved quality assurance program is in place must be qualified for use as AK. Title 40 CFR 194.22(b) lists four methods that may be used individually or in combination to qualify such data: 1) peer review, 2) use of corroborative evidence, 3) confirmation by measurements, or 4) qualification of previous quality assurance programs. DOE is proposing to limit confirmatory measurements only to those cases where historical records are incomplete. DOE proposes not to confirm the newly generated information for 95 percent of the waste since this information will be collected under a certified quality assurance characterization program. The specific needs for waste analysis are described in DOE's proposed characterization plan as data quality objectives and quality assurance objectives.

4.3 DATA QUALITY OBJECTIVES AND QUALITY ASSURANCE OBJECTIVES

Data quality objectives are proposed by the implementer, in this case DOE, to define what information is necessary to characterize waste. They are qualitative and quantitative statements that clarify program's objectives, define appropriate types of data, and specify tolerable decision error rates that will be used as the basis for establishing the quality and quantity of data needed to support decisions.

The EPA document SW-846 "Test Methods for Evaluating Solid Waste Physical/Chemical Methods" defines data quality objectives as:

> "Data quality objectives…for the data collection activity describe the overall level of uncertainty that a decision-maker is willing to accept in results derived from environmental data" (EPA, 2002; pages 1–2).

Quality assurance objectives establish minimum requirements for the measurement and representation of data. Quality assurance objectives are designed to provide information that will satisfy the data quality objectives. They correspond to EPA's "measurement quality objectives." The elements of the quality assurance objectives are:

- Data accuracy: the degree to which data agree with an accepted reference or true value.
- Data precision: a measure of the mutual agreement between comparable data gathered or developed under similar conditions expressed in terms of a standard deviation.
- Data representativeness: the degree to which data accurately and precisely represent a characteristic of a population, a parameter, variations at a sampling point, or environmental conditions.
- Data completeness: a measure of the amount of valid data obtained compared to the amount that was expected.
- Data comparability: a measure of the confidence with which one data set can be compared to another [40 CFR 194.22(c)].

Documents 1 and 2 describe the data quality objectives and quality assurance objectives to meet the requirements in WIPP's EPA Certification and RCRA Permit, respectively.

4.4 DOE'S PROPOSED CHARACTERIZATION PLAN TO ADDRESS EPA REQUIREMENTS

The characterization requirements in the EPA Certification, addressing also the requirements in the Land Withdrawal Act, are described in Chapter 3. The data quality objectives, characterization, and implementation methods to address these requirements are provided in Document 1 and summarized in Table 4.1. The characterization methods to meet EPA requirements are the following:

- Acceptable Knowledge: the type of information used to define waste streams is described in Sidebar 2.1. For the RH-TRU waste characterization program, 95 percent of the waste will be packaged or repackaged during its preparation for shipment to the WIPP. Information collected under these packaging and repackaging programs is considered to be AK information along with historical data.
- Dose-to-Curie Conversion: the curie content of waste is based on a dose rate measurement taken with a calibrated field instrument. The dose rate measurement is associated with documented isotopic distributions in the waste through the use of empirically developed conversion factors. Dose-to-curie estimates may also be based on the measurement of an indicator activity (e.g., cesium-137) and correlation of other radionuclides by isotopic conversion factors. This method requires good knowledge of the isotopic distribution in the waste.
- Visual Examination: this method involves the removal of items from the container and their inspection and identification. The examination will be recorded on a signed data form accompanied by visual evidence such as an audio or video tape. Visual examination only provides a visual image of the content of waste containers. It does not provide any information on the isotopic composition of waste.
- Radiography: this method involves the use of penetrating radiation to investigate the contents of containers. The examination will be recorded on a signed data form

accompanied by visual evidence such as an audio or video tape or other media. Radiography only provides an image of the content of waste containers. It does not provide any information on the isotopic composition of waste.
- Characterization at the Time of Packaging: this method involves visual examination and other characterization methods applied as waste is packaged for shipment to WIPP. DOE distinguishes this method from visual examination for the following reasons:

1. it applies to 100 percent of the waste in the waste stream;
2. it generally focuses on the development or collection of AK information as opposed to the confirmation of such information;
3. it applies to waste that is not in its final shipping container; and
4. it may require the use of other characterization methods (i.e., AK, visual examination, or radiography) to determine the contents of smaller containers that cannot be readily characterized without testing.[2]

- Direct Assay: this method involves using radioassay systems that are qualified for CH-TRU waste characterization.
- Counting containers: this method involves automatically counting containers to maintain the record of the amount of metals introduced in the repository.

4.5 DOE'S PROPOSED CHARACTERIZATION PLAN TO ADDRESS NMED REQUIREMENTS

The characterization requirements in the current RCRA Permit are described in Chapter 3. The proposed data quality objectives and characterization methods to address these characterization requirements are presented in Document 2 and summarized in Table 4.2.

Characterization methods and the relative quality assurance objectives of precision, accuracy, representativeness, completeness, and comparability are discussed in Attachment R of Document 2. The same characterization methods have been discussed for Document 1 (see above). Radiography, along with other non-destructive examination and non-destructive assay methods, is also briefly described in Appendix G.

---

[2]The committee observes that "characterization at the time of packaging" may cause a semantic confusion with AK, which is usually associated with historical information. Characterization at the time of packaging generates new information through visual examination, radiography, or other characterization methods; it may also include historical data on a particular waste stream.

TABLE 4.1 DOE's Data Quality Objectives, Characterization and Implementation Methods to Address EPA Requirements

No.	Data Quality Objectives	Characterization Method	Implementation Methods
1	Account for TRU activity. Rationale: This requirement is related to the LWA's definition of TRU waste and to 40 CFR 191. The purpose for this data quality objective is twofold: 1) ensure that waste is TRU (TRU activity equal to or greater than 100 nanocuries per gram of waste); and 2) determine the plutonium-239 content to satisfy the release limits set forth in 40 CFR 191.	1. AK method 2. Dose to curie 3. Direct assay 4. Characterization at the time of packaging[1]	Use AK to determine the relationship of TRU activity to total activity (see no. 6 below) for each waste stream (confirmation unnecessary). AK must include TRU information (sufficient to demonstrate the waste stream concentration exceeds 100 nanocuries per gram), or the waste stream value must be established by sampling and measurement, or the TRU activity must be measured for each container.
2	Ensure that waste is of defense origin. Rationale: The LWA allows only disposal of defense-related transuranic waste in WIPP.	1. AK method	Use AK to determine whether waste was generated in a defense-related activity.
3	Ensure waste is RH (surface dose equal to or greater than 200 mrem per hour). Rationale: operational requirement to determine handling and characterization requirements.	1. Standard industry survey methods 2. Characterization at the time of packaging	Use standard industry survey methods to measure and report surface dose rate. Survey instrument calibrations must be traceable by the National Institute of Standards and Technology.
4	Account for total activity. Rationale: this requirement comes from the LWA limit of 5.1 million curie of total activity for RH-TRU waste.	1. AK method 2. Dose to curie 3. Direct assay 4. Characterization at the time of packaging	Use AK to determine total activity of the waste stream (confirmation unnecessary), or establish a waste stream value by sampling and measurement, or measure the total activity of each container.
5	Limit canister activity to less than 23 curies per liter. Rationale: this requirement comes from the LWA limit on canister activity.	1. AK method 2. Dose to curie 3. Direct assay 4. Characterization at the time of packaging	Use AK to evaluate each waste stream and show that this limit cannot feasibly be approached for that particular waste stream. For the rare exception where AK indicates that 23 curies per liter is hypothetically feasible, measurements of Total Activity should be obtained on few representative containers to establish a waste stream value, or a site may choose to measure the total activity of each container.

6	Limit surface dose rate of each container to less than 1,000 rem per hour. Rationale: This is a LWA requirement.	1. Standard industry survey methods 2. Characterization at the time of packaging	Survey instrument calibrations must be traceable by the National Institute of Standards and Technology.
7	Limit WIPP inventory to 5 percent by volume of canisters with a dose rate greater than 100 rem per hour. Rationale: This is a LWA requirement.	1. Standard industry survey methods 2. Characterization at the time of packaging	Use standard industry survey methods to measure and report Surface Dose Rate. Survey instrument calibrations should be traceable by the National Institute of Standards and Technology.
8	Account for ferrous and non-ferrous metals. Rationale: this is a requirement introduced in the EPA's Certification.	1. Count containers	WWIS [WIPP Waste Information System] electronically counts waste containers. Confirmation unnecessary.
9	Account for cellulosics, plastics, rubber. Rationale: this is a requirement introduced in the EPA Certification	1. AK method 2. Visual examination 3. Characterization at the time of packaging	Use AK to determine Summary Category Group (SCG) of each waste stream. Use generator/storage site SCG designation from site documents. WWIS electronically calculates and assigns for each emplaced container (0 for homogenous solids and soils/gravel, and 50 percent of container weight for debris). Confirmation unnecessary if AK information is complete.
10	Account for free water (up to 1 percent of free water by volume). Rationale: WIPP was certified on the basis of a performance calculation using 1 percent of water by volume in the waste. This limit originates from a transportation requirement.	1. AK method 2. NDE such as Radiography 3. Visual examination 4. Characterization at the time of packaging	If AK documents the use of liquid management procedures (e.g., removal or absorption), or a process that precludes liquids (e.g., material input records or thermal treatment), then assign 0 percent of liquid by volume for each waste stream. Otherwise assign 1 percent volume to the entire waste stream. Confirmation unnecessary. When AK is lacking, use a non-destructive examination, visual examination, or characterization at the time of packaging on a sampling basis to show that a waste stream contains less than one percent residual liquids.

NOTE: AK=acceptable knowledge; LWA=Land Withdrawal Act, NDE=non-destructive examination. Information in this table is presented in Document 1 and supplemented with additional information provided to the committee.

[1]Characterization at the time of packaging is defined in Document 1 as a characterization method. However, this encompasses several activities, such as visual examination, chemical assays, as explained in the description of characterization methods.

SOURCE: DOE (DOE-CBFO, 2002a; 2002c).

TABLE 4.2 DOE's Data Quality Objectives, Characterization and Implementation Methods to Address NMED Requirements

No.	Data Quality Objectives	Characterization Method	Implementation Method
1	Assign hazardous waste codes. Rationale: RCRA requirement as it applies to WIPP.	1. AK method 2. Tests on waste samples	Use AK to delineate waste streams, assign hazardous waste codes and summary category groups (SCG) to waste streams, and assign individual containers to waste streams. Use generator/storage site records of container assignments to the waste stream. Confirmation unnecessary. If AK is insufficient, perform tests on a representative (10 percent) sample of the waste.
2	Identify Physical Form. Rationale: RCRA requirement as it applies to WIPP.	1. AK method 2. Visual examination 3. NDE such as radiography	Summary category groups define the physical form of the waste as solidified solids (SCG=S3000), soil/gravel (SCG=S4000), or debris (SCG=S5000). Confirmation is unnecessary. If AK is insufficient, perform tests such as VE and NDE on a representative (10 percent) sample of the waste.
3	Limit residual liquids to less than 1 percent volume of RH canister (or drum for 160B). Rationale: residual liquids limit arises from transportation and operational safety considerations. Note that the rationale is different from that in Table 4.1 item 10. Residual liquids are a prohibited item in the current RCRA Permit.	1. AK method 2. Visual examination 3. NDE such as radiography	Use AK to demonstrate that the waste stream does not contain residual liquid, or that individual containers contain less than 1 percent by volume of residual liquid (confirmation unnecessary), or sample and NDE/VE to determine the waste stream does not contain residual liquid, or NDE/VE each container in a waste stream. AK must document the use of liquid management procedures (e.g., removal or absorption), or a process that precludes liquids (e.g., material input records or thermal treatment), or there must be a record for each container indicating less than 1 percent by volume of residual liquid, or NDE/VE of a few representative containers (10 percent[1]) from the waste stream can show the absence of residual liquids, or every container must undergo NDE/VE.
4	Limit residual polychlorinated biphenyls to less than 50 ppm. Rationale: polychlorinated biphenyls are a prohibited item in RCRA Permit if present in a concentration greater than or equal to 50 ppm.	1. AK method 2. Visual examination 3. NDE such as radiography	Use AK to determine the absence of possible sources of polychlorinated biphenyls such as transformer oil. If AK is insufficient, perform tests such as VE and NDE on a representative (10 percent) sample of the waste. VE and NDE are used, for instance, to look for parts of or entire transformers.
5	Exclude additional prohibited items	Use EPA hazardous waste	Additional prohibited items listed in the RCRA Permit (see

	(such as pyrophorics, incompatible chemicals, explosives, compressed gases, ignitable, corrosive, and/or reactive wastes). Rationale: these items are prohibited in the current RCRA Permit.	codes	Chapter 3) are addressed by the EPA hazardous waste codes assigned to an RH-TRU waste stream. RH-TRU waste containing any one of these items would have EPA hazardous waste codes that would disqualify it from management, storage, and disposal under the WIPP Part A requirements.
6	Identify and quantify volatile organic compounds. Rationale: any waste container that does not have volatile organic compounds concentrations reported for the headspace is currently a prohibited item in the RCRA Permit.	No volatile organic compounds measurement proposed on RH-TRU waste.	Use bounding analysis for RH-TRU waste contribution to reduce current RCRA Permit Room Limits.

NOTE: AK=acceptable knowledge; NDE/VE=non-destructive and visual examination (see Appendix G); VE=visual examination. Information in this table is presented in Document 2 and supplemented with additional information provided to the committee.
[1]See committee's Finding 2A, Example #6 in Chapter 5. Other elements of this table are also discussed in Chapter 5.
SOURCE: DOE (DOE-CBFO, 2002b; 2002c).

4.6 MAJOR DIFFERENCES BETWEEN THE JULY 2001 AND THE MARCH 2002 DRAFTS

As noted in Chapter 1, the committee reviewed two drafts of the proposed characterization plan. This section outlines the five major differences between the two drafts, as follows:

- *Change in the approach to the determination of prohibited items in Document 2.* The determination of the absence of all the prohibited items listed in the RCRA Permit was part of the requirements in the July 2001 draft of Document 2 (DOE-CBFO, 2001b; Item 2; pages 2–27). In the March 2002 draft, the only prohibited items explicitly listed as data quality objectives are residual liquids (if more than 1 percent by volume of waste) and polychlorinated biphenyls (if equal to or greater than 50 ppm). Remaining incompatible prohibited items (ignitable, corrosive, reactive, pyrophoric and incompatible waste, explosives, and compressed gases) will be addressed by excluding from shipment hazardous waste with codes corresponding to ignitable (D001), corrosive (D002), and reactive waste (D003). Pyrophoric materials, explosives and compressed gases fall under the definition of reactive waste (D003). Other waste prohibited in WIPP, such as explosives and compressed gases, can be excluded through hazardous waste code screening because any material whose code does not appear in the list of hazardous waste codes allowed in WIPP cannot be shipped. This drives the determination of the remaining prohibited items from a container-by-container basis to a waste stream level, where hazardous waste code assignment decisions are typically made. The hazardous waste code(s) determination will be accomplished using AK. According to DOE, the 1 percent free liquid criterion was retained as a RCRA operational compliance consideration rather than for performance reasons (see Appendix D).
- *Definition of a representative sample of containers to perform additional tests when AK is insufficient.* The March 2002 draft defines for the first time a representative sample as "at least 10 percent of the samples," selected by the generator site (DOE-CBFO, 2002a; Attachment B; page 26). The committee discusses the issue of 10 percent representative sample size in Chapter 5, Finding 2A, Example #6.
- *Change of quality assurance objective to determine the total activity of waste by direct assay.* The July 2001 draft of Document 1 read:

  "Track the RH-TRU waste total activity inventory by the quantification of total activity for a unit (waste stream or individual container) within a factor of five of the true value with a confidence level of 95 percent" (DOE-CBFO, 2001a; Attachment A; page 10).

  In the March 2002 draft of Document 1, this quality assurance objective was replaced by the quality assurance objective used for CH-TRU waste (DOE-CBFO, 1999b, Appendix A). Moreover, each site will demonstrate the ability of its direct assay measurement system to meet the criteria but will not be subject to the CH-TRU waste performance demonstration program.
- *Distinction between visual examination and characterization at the time of packaging.* Most of the RH-TRU waste will either be generated or repackaged. DOE makes a distinction between the characterization at the time of packaging and visual examination (DOE-CBFO, 2002a; Attachment B; pages 26 and 29). The rationale for the distinction is given above (see also the definition of the method "characterization at the time of packaging" in Section 4.4).

- *More detailed description of characterization requirements and RH-TRU waste inventory in both documents.* The March 2002 draft reports data quality objectives and quality assurance objectives for both EPA and RCRA Permit requirements. The documents also present additional information about generator sites and RH-TRU waste inventories.

While CH- and RH-TRU waste share the same characterization requirements described in WIPP's RCRA Permit and EPA Certification, DOE's approach to meet these requirements for RH-TRU waste is quite different from the way the are met for CH-TRU waste, as explained below.

4.7 COMPARISON BETWEEN THE CH- AND RH-TRU WASTE CHARACTERIZATION PLANS

The CH-TRU waste characterization plan approved by EPA and NMED is described in Appendix F. Briefly, the characterization plan for CH-TRU waste is based on a 100 percent confirmation of AK information collected on each waste stream. Confirmation is accomplished by radioassay, headspace gas sampling, radiography, or visual examination. For homogeneous wastes,[3] this includes taking core samples of a fraction of waste containers prior to their shipment to WIPP. For debris wastes, a 100 percent confirmation program (with headspace gas and radiography determinations for RCRA compliance, and non-destructive assay and radiography determinations for EPA's compliance) was negotiated between DOE and NMED. Rather than proposing the same approach as for CH-TRU waste, DOE proposes a "performance-based" approach to characterize RH-TRU waste because of the high radiation fields and the negligible impact of this waste on the performance of the repository.

The most significant difference between the CH- and the proposed RH-TRU waste characterization plans is that the latter does not require confirmatory testing, sampling, or analysis on 100 percent of containers and allows radiography and visual examination only on a representative selection of containers. DOE proposes to use AK whenever possible to accumulate the information on RH-TRU waste and to perform confirmation measurements only when the information collected on the waste stream is not qualified as AK. The rationale provided by DOE in Document 2 relies on the 95 percent of RH-TRU waste volume that is to be generated or repackaged. Instead of confirming AK, generator sites must provide assurances that the waste is adequately characterized at the time of packaging.

On the basis of the Sandia impact analyses, DOE determined that the proposed RH-TRU characterization plan does not require the determination of material parameter category weights and headspace gas analysis, whereas they are required in the CH-TRU waste characterization plan. Instead, DOE proposes to use bounding values for these parameters for RH-TRU waste. Concerning the WIPP's limit on volatile organic compound emissions, the characterization plan for CH-TRU waste requires tracking individual waste container headspace gas concentrations, which are determined by headspace gas sampling and analysis of CH-TRU waste containers.

There is no characterization objective for a direct measurement of the headspace gases in the proposed RH-TRU waste characterization plan because canisters will be located in the walls of the disposal rooms behind shield plugs. Calculations have shown

---

[3]Homogeneous waste is one of the "Summary Category Groups" that indicate the final form of the waste. Homogeneous waste is waste consisting of one main constituent (for instance, sludge) as opposed to the other two category groups: soils and gravel, and debris waste.

that it is possible to account for volatile organic compounds emission rates from RH-TRU waste indirectly.[4] This is accomplished by conservatively incorporating the maximum potential volatile organic compounds contribution from RH-TRU waste in the maximum allowable volatile organic compounds emission limits established by the RCRA Permit (Spangler et al., 2002).

---

[4]To the best of committee's knowledge, experiments on volatile organic compound emissions from RH-TRU waste have not been conducted.

5

Assessment of the Proposed Characterization Plan

This chapter provides the committee's assessment of DOE's proposed plan for the characterization of RH-TRU waste. On June 28, 2002 DOE submitted the characterization plan to EPA and NMED. The committee did not review the plan as submitted; therefore, findings and recommendations in this chapter apply only to the March 2002 draft.[1]

According to the statement of task, findings and recommendations are organized as follows:

1. context of RH-TRU waste characterization;
2. characterization plan's technical soundness;
3. protection of worker safety and health; and
4. compliance with regulatory requirements.

This assessment is based on technical considerations and on the criteria listed in the statement of task (Sidebar P.1).

5.1 CONTEXT OF REMOTE-HANDLED TRANSURANIC WASTE CHARACTERIZATION

Below are some general findings about the DOE's RH-TRU waste inventories. According to the information gathered, the RH-TRU waste volume estimated for shipment to WIPP will be about 3,800 cubic meters or 2 percent of the total TRU waste volume allowed in the repository. Even if the information presented in DOE inventories were not accurate, the RH-TRU waste emplaced in WIPP cannot exceed the limits for RH-TRU waste established in the Land Withdrawal Act: 7,080 cubic meters (about 4 percent of the total allowed TRU volume) and 5.1 million curies.

While RH-TRU waste volume represents a small fraction of the allowed TRU waste in WIPP, it accounts for 14 percent of the total curie activity in DOE's TRU inventory. However, the number of TRU curies in the RH-TRU waste inventory is 2 orders of magnitude lower than that in the CH-TRU waste inventory expected in WIPP (see Table 2.3 in Chapter 2). Also, most of the total (CH plus RH) long-term (i.e., TRU) activity expected in WIPP comes from CH-TRU waste, with only a 0.5 percent contribution from RH-TRU waste.

According to the performance assessment calculations by Sandia National Laboratories, the inventory of RH-TRU waste expected in WIPP will have a negligible

---

[1] The committee also reviewed a previous draft (July 2001), which was the object of its interim report (see excerpt in Appendix C).

impact on the repository's compliance with EPA radiological regulatory standards, compared to CH-TRU waste.

The committee acknowledges that there are uncertainties in DOE's volume and radioactivity inventories of RH-TRU waste. Also, the committee did not assess the validity of the Sandia impact analyses. However, the committee notes that these performance assessment calculations were performed using a peer-reviewed and EPA-approved performance assessment tool, which was the basis for WIPP certification. The committee recognizes that there may be uncertainties associated with the results of these impact analyses and that there may be scenarios that require a new set of performance assessment calculations.[2] In response to the committee's question on uncertainties, Sandia National Laboratories reported that:

> "Uncertainties in Performance Assessment are dominated by those associated with geologic processes, chemical interactions of the waste, and possible human activities in the future. For example, the intrinsic permeability of the proximal host rock can vary over seven orders of magnitude; uncertainties with respect to microbial degradation of organic materials affect several key intermediate variables, such as long-term repository pressure and saturation; assumptions regarding human activities introduce the most notable impacts on repository performance. With these uncertainties in mind, it can be stated that any additional uncertainty associated with waste components within the currently estimated RH-TRU inventory are not significant to long-term repository performance" (Knowles, 2002).

These considerations, as well as the findings and recommendations in this section, set the context for RH-TRU waste characterization. This context is different than that of CH-TRU waste characterization because of the differences in volumes and in isotopic composition, as explained in Chapter 2.

Finding 1 A: According to DOE inventories, 95 percent of the RH-TRU waste to be disposed of in WIPP has yet to be generated or needs to be processed, packaged, or repackaged. This waste will be characterized (through visual examination and physical and chemical analyses) at the time of packaging. **Recommendation: DOE should emphasize the argument that the characterization information collected for most of RH-TRU waste does not need confirmatory measurements because the repackaging or generation of waste will be carried out under a certified quality assurance program.**

Rationale: According to the information provided by DOE, over 95 percent of the RH-TRU waste inventory (to be generated, to be packaged or repackaged waste in Figure 2.2) will be characterized using visual examination[3] and physical and chemical methods. This leaves a small fraction of RH-TRU waste with more limited historical, analytical, radioassay, and visual examination information or which was not

---

[2] An unexpected brine saturation of the repository in less than 300 years, a change in the scope of the WIPP facility, or a change in the configuration of underground disposal are examples of scenarios for which a new set of performance assessment calculations for both CH- and RH-TRU waste would be warranted.

[3] The committee notes that visual examination only provides an image of the content of waste containers. It does not provide any information on the isotopic composition of waste.

characterized under a certified quality assurance program; in this case, AK may require confirmation.

Therefore, for over 95 percent of RH-TRU waste, confirmatory measurements may not be necessary as long as this waste is generated or repackaged using a certified quality assurance program plan. If the volume of RH-TRU waste represents between 2 and 4 percent of the total inventory of TRU waste (176,000 cubic meters), and the information collected for over 95 percent of RH-TRU waste does not need confirmation, then only between 0.1 and 0.2 percent of the total TRU inventory needs confirmatory activities for AK. This is an important observation and it could have a substantial impact on characterization programs proposed for RH-TRU waste.

Finding 1B: For 95 percent of the RH-TRU waste inventory, AK refers to newly generated information acquired during waste generation and packaging. **Recommendation: DOE should utilize a different term than "AK" for this newly generated information.**

Rationale: This newly generated information does not include only historical information on the waste. It can include, for instance, analytical data, radioassay data, and visual examination records obtained during waste generation or repackaging. DOE considers this newly generated information, which may not require confirmation, as AK. The committee believes that this is a potentially confusing terminology because AK is usually associated with historical information, which requires some confirmation (see also Sidebar 2.1).

The committee acknowledges that visual examination during RH-TRU waste generation or repackaging does not provide information on the radiological content or on the presence of corrosive, reactive, or chemically hazardous waste. However, visual examination is not the only characterization method proposed for newly generated or repackaged waste (see description of the "characterization at the time of packaging" method in Chapter 4, Section 4.4).

5.2 CHARACTERIZATION PLAN'S TECHNICAL SOUNDNESS

DOE's stated objective is to propose a performance-based characterization plan for RH-TRU waste. The committee interprets "performance-based characterization plan" as a plan ensuring repository integrity, protection of the environment, and health and safety of the public and workers while keeping in consideration tradeoffs between greater characterization accuracy and worker risks (see Chapter 4). This section addresses the characterization activities that DOE proposes to implement the data quality objectives listed in Tables 4.1 and 4.2.

Finding 2A: The committee found that several characterization activities are based on non-technical considerations. Therefore, this characterization plan is not completely performance based. The committee questions the technical basis of some of the characterization activities proposed by DOE with respect to its stated goal of adopting a performance-based approach. **Recommendation: The committee acknowledges that DOE must consider many non-technical factors in composing its characterization plan. However, DOE should propose only characterization activities that have a technical, health and safety, or regulatory basis.**

Rationale: DOE itself recognizes that the characterization plan is based on various considerations other than science and technology. In its response to the findings in the committee's interim report, DOE stated:

> "Nearly all resulting characterization objectives are unrelated to repository performance or to safety/technical considerations" (see Appendix D).

and, during a recent conference on waste management, DOE explains:

> "Based on the results of repository modeling...the DOE believes there are no specific RH TRU waste parameters that need to be measured with precision in order to assure repository integrity and assure protection of human health and the environment. This does not mean that no characterization of RH TRU waste is necessary. To the contrary, in order to meet the requirements of the facility design and the facility waste acceptance criteria, specific needs for waste analysis have been identified" (Gist, et al., 2002; pages 2–3).

The facility design and the waste acceptance criteria address the requirements in the EPA Certification and RCRA Permit. Regulatory drivers are listed in Chapter 3 and other non-technical considerations at the beginning of Chapter 4. Most of the characterization activities proposed by DOE do not have a technical basis but arise from considerations aimed at maintaining an effective working relationship among DOE, EPA, and NMED. The committee acknowledges that maintaining an effective working relationship among DOE and WIPP's regulators is important; however, DOE should propose only characterization activities that correspond to technical, health and safety, or regulatory objectives.

Concerning the technical basis of some of the characterization activities in Documents 1 and 2, the Sandia impact analyses show that radioisotope, free liquids, cellulosics, plastics, rubber, and ferrous metal content in the RH-TRU waste inventory do not have an impact on repository performance (see Chapter 4).

In spite of these impact analyses, Documents 1 and 2 still propose characterization activities to determine the above parameters in RH-TRU waste. The committee provides below examples of characterization activities that do not appear to have a technical basis in the context of RH-TRU waste characterization.

Example #1: DOE justifies the characterization requirement concerning free water or liquid content for both EPA and NMED as a transportation-related requirement or as a requirement present in the WIPP Safety Analysis Report.[5] DOE provides the rationale for this characterization requirement as follows:

> *"[EPA Characterization Objectives] Account for Free Water,* is a regulatory expectation that the RH-TRU Team believes is prudent to meet even though no amount of free water in the RH-TRU waste will impact repository performance. Note that Appendix WCL.5 [of the compliance certification application] states that, 'Consequently, there is no need to monitor water in the waste for

---

[5]For a definition of Safety Analysis Report see the glossary, Appendix I.

[4]An unexpected brine saturation of the repository in less than 300 years, a change in the scope of the WIPP facility, or a change in the configuration of underground disposal are examples of scenarios for which a new set of performance assessment calculations for both CH and RH-TRU waste would be warranted.

compliance with 40 CFR § 194.24(c)' The RH-TRU Team recommends that free water in RH-TRU waste be accounted for by some means.

[NMED Characterization Objectives] Limit residual liquids to [less than] 1 [percent] volume of RH canister (or drum for 160B), is related to a regulatory requirement to provide secondary containment at the WIPP for potential spills. An analysis justifying the 1 [percent] for CH-TRU waste at the WIPP is part of the HWFP [Hazardous Waste Facility Permit] record, however a similar analysis has not been done for RH-TRU waste. The RH-TRU Team believes it is prudent to include this limit as an RH-TRU waste requirement also" (see Appendix D).

DOE acknowledges that the free water/liquid requirement is an operational concern that does not affect the long-term performance of WIPP. Given the results of the Sandia impact analyses, the committee questions the technical basis of this characterization activity.

Example #2: One of the data quality objectives listed in Document 1 is the determination of TRU activity (expressed as plutonium-239 content) of RH-TRU waste to comply with the EPA radionuclide release limits. The Sandia impact analyses showed that, even if the entire RH-TRU waste radionuclide inventory were composed of plutonium-239, WIPP would still be in compliance with 40 CFR 191 because of the small volume of RH-TRU waste to be emplaced in WIPP. Therefore, the actual measurement of TRU activity does not have a technical basis for RH-TRU waste. The determination of TRU activity could be addressed by assigning a conservative boundary value of plutonium-239 representing the total RH-TRU waste inventory.

Example #3: In the EPA compliance certification application, DOE established limits for ferrous metal content of waste to be emplaced in WIPP. The rationale is the following:

"Ferrous and ferrous-alloy metals (and their corroded products) provide the reactants that reduce radionuclides to lower and less-soluble oxidation states. As discussed in Appendix WCA [Waste Characterization Analysis in the compliance certification application], the anticipated quantity of these metals to be emplaced in WIPP is two to three orders of magnitude in excess of the quantity required to assure reducing conditions. The waste containers supply more than enough iron to provide adequate reductant. Therefore, no upper or lower limit need be established for the quantity of ferrous and ferrous-alloy metals that may be emplaced, beyond the present projection of containers" (DOE, 1996a; Appendix WCL.2).

According to the information gathered, the minimum amount of ferrous metals will be satisfied by CH-TRU waste containers. The committee was unable to determine the reason for including the implementation of this characterization objective in the RH-TRU waste characterization plan, even though the method (counting containers) is a simple practice performed during waste packaging.

Example #4: The Sandia impact analyses showed that, even if the entire RH-TRU waste inventory were composed of cellulosics, plastics, and rubber, it would only have a negligible impact on the repository performance because of the small volume of RH-TRU waste emplaced. Yet, DOE proposes the implementation of this characterization objective in Document 1. Although this objective is not performance based in the case of RH-TRU waste, the committee acknowledges that DOE does not propose to actually measure cellulosics, plastics, and rubber in the waste but uses a conservative estimate described in Table 4.1.

Example #5: The determination of prohibited items (see Section 3.3, Chapter 3) listed in the RCRA Permit may not be appropriate for RH-TRU waste when the risks associated with hazardous chemical waste are balanced against radiological risks and costs associated with characterization. According to the information gathered, the prohibited items' determination represents one of the major difficulties in the characterization of RH-TRU waste. In DOE's characterization plan, there is no analysis of the health and safety implications of the prohibited items in RH-TRU waste. It is important to analyze these yet-undefined safety implications and balance them against potential radiological risks to workers and associated costs of identifying prohibited items in RH-TRU waste. Such an analysis would be helpful to DOE in comparing workers risks and the associated characterization costs. The results of the analysis could support and strengthen DOE's RH-TRU waste characterization activities with respect to prohibited items.

DOE's proposed characterization plan was significantly modified between July 2001 and March 2002 concerning the determination of prohibited items in RH-TRU waste. In the latter draft, DOE proposes to determine the absence of only free liquids and polychlorinated biphenyls (within their limits). The remaining prohibited items listed in the RCRA Permit (ignitable, corrosive, reactive, pyrophoric and incompatible waste, explosives, and compressed gases) are determined using the EPA hazardous waste codes. This determination is accomplished using AK on a waste stream basis.

This change drives the prohibited items determination (except for residual liquids and polychlorinated biphenyls) from a container-by-container basis to a waste stream basis where hazardous waste code assignment decisions are typically made. The committee supports this change since it streamlines characterization operations but still addresses the prohibited items objective in the RCRA Permit.

The committee could not determine the technical basis for including the determination of polychlorinated biphenyls (over 50 ppm) in the context of RH-TRU waste. DOE provided the following rationale:

> "Polychlorinated biphenyls are subject to EPA regulation under TSCA [Toxic Substances Control Act] and are regulated regardless of whether or not they are included in a RCRA permit as a prohibition" (see Appendix D).

In the case of a change in the polychlorinated biphenyl disposal regulations,[6] DOE must submit a request for a change in the RCRA Permit to remove the 50 ppm provision. This would not be necessary if this provision were not explicitly mentioned in the RCRA permit.

Example #6: The March 2002 draft indicates that, if AK data are not sufficient, then 10 percent of waste should be characterized with alternative methods (see Tables 4.1 and 4.2). The committee could not determine the technical basis for choosing 10 percent of waste as a representative sample of the waste stream. In general, the size of a representative sample is the result of an analysis involving statistical considerations as

---

[6]The new polychlorinated biphenyls disposal regulations (Disposal of Polychlorinated Biphenyls (PCBs), 63 Federal Register 35384, effective August 28, 1998) allows polychlorinated biphenyl-contaminated radioactive waste to be disposed of without a TSCA permit provided the waste meets the requirements for disposal in a non-hazardous waste landfill or a hazardous waste landfill. This recent regulatory change would allow disposal at WIPP of some polychlorinated biphenyls-contaminated wastes where the polychlorinated biphenyls component exceeds 50 ppm. On March 22, 2002 DOE submitted an "Initial Report" to EPA Region VI requesting authorization to dispose of polychlorinated biphenyls in WIPP.

well as waste stream profile considerations. No such analysis or any other justification was provided in the March 2002 draft. A 10 percent waste sample size may be too big or too small, depending on the waste stream, on the quality of AK available, and on the tolerable decision error rates (see Finding 2C).

Finding 2B: The requirements to qualify information collected on each waste stream, whether by AK or by any other method described in 40 CFR 194.22(b), have not been established with specificity in the submittal documents. **Recommendation: In the site-specific accompanying documents, DOE should present clear and technically defensible data qualification requirements for its RH-TRU waste characterization plan.**

Rationale: The committee recommends that DOE follow a more structured approach in the development of its characterization plan. A structured approach, such as that provided by defining data quality objectives and quality assurance objectives, would clearly define the data needed for characterization, identify the basis for the data need, and then logically determine how to meet those needs. The committee does not expect quantitative quality assurance objectives, especially for AK information. However, it is important to discuss how to determine when or whether AK could meet data quality objectives and under which conditions it would be necessary to supplement AK using other characterization methods (e.g., dose to curie conversion, visual examination, radiography, time of packaging methods, direct assay, surface dose rate). This recommendation is also consistent with the recommendation of a second peer review on DOE's characterization plan.[7]

According to the information gathered, DOE will submit to EPA and NMED along with the draft characterization plan, a sample of completed characterization documents for three RH-TRU waste generator sites (Energy Technology Engineering Center, Battelle Columbus Laboratories, and Oak Ridge National Laboratory). The committee supports this initiative. These companion documents to the characterization plan should be useful in clarifying the intended methods of implementation. The committee could not review these methods, given that these documents were not available at the time this report was written.

Finding 2C: DOE's proposed characterization plan does not adequately address the issue of tolerable decision error rates associated with all characterization information. **Recommendation: DOE's proposed characterization plan should address tolerable decision error rates associated with characterization information. These errors should not be overly stringent so as to negatively impact the sites' ability to implement ALARA.**

Rationale: Tolerable decision error rates are never clearly defined in Document 1 and Document 2. Tolerable decision error rates are determined by weighing the consequences of mischaracterization against the costs (including worker risks) of achieving better characterization. Once established, tolerable decision error rates can be

---

[7]The Institute for Regulatory Sciences, which reviewed the July 2001 draft of DOE's proposed characterization plan, recommended the following: "A detailed procedure for determining whether there is sufficient AK available on a waste, should be developed as part of the permit application. This procedure should be consistent across all waste generating sites. […] [A] detailed procedure should be provided to go to other characterization methods if AK is found to be insufficient" (Institute for Regulatory Sciences, 2001; page 77).

used to identify the attendant quality assurance requirements for sampling (i.e., the measurement quality objectives, which DOE calls quality assurance objectives). Tolerable decision error rates may not be specified in laws and regulations, but they are a critical judgment that an implementer should propose when interpreting legal or regulatory limits to address the uncertainties inherent in characterization data. As an illustration, Document 1 reads as follows:

> "Tolerable decision error: The limit on residual liquids has been specified with no associated error; therefore, any container that contains more than 1 percent by volume residual liquids cannot be shipped to WIPP" (DOE-CBFO, 2002a; Attachment B; page 7).

The tolerable decision error (and its rate) in the measurement of the liquid volume in a container is never clearly defined. More importantly, the above quote from Document 1 seems to suggest that there is a zero tolerable decision error (and a zero tolerable decision error rate) on the determination of residual liquids. Another example of unclear tolerable decision error rate concerns the quantification of the total activity. The March 2002 draft proposes to use the same quality assurance objective for the total activity of waste as the one used for CH-TRU waste (see Chapter 4, Section 4.6). However, the committee still cannot determine with clarity what is tolerable decision error rate associated with the measurement of the total activity.

It is premature for DOE to develop quality assurance objectives until it has proposed tolerable decision error rates that limit risks related to the waste, and that recognize potential tradeoffs in costs and worker risks. Without setting tolerable decision error rates to create a bridge between the data quality objectives and the characterization methods, it will be impossible to determine whether the waste characterization plan requires too much or too little in terms of the quality of the data needed. The quality assurance objectives should be presented in the site-specific implementation plans and could vary from site to site. The committee supports DOE's plan to submit to EPA and NMED the three site-specific (Energy Technology Engineering Center, Battelle Columbus Laboratories, and Oak Ridge National Laboratory) implementation plans.

It is important to recognize that tolerable decision error rates will have an impact on characterization operations at the generator sites. The ALARA principle must be applied at DOE facilities according to the Code of Federal Regulations (10 CFR 835.101). Therefore, the proposed characterization plan should not set forth overly stringent requirements (such as zero tolerable decision error rates) that could negatively impact the sites' ability to implement ALARA (see also Finding 3A).

Finding 2D: It is not clear how visual examination and radiography can confirm AK information for prohibited items. For example, visual examination and radiography cannot distinguish between corrosive and non-corrosive liquids, whereas AK may provide records of the existence of such liquids in the waste. Historical AK may be a better indicator of some of the currently prohibited items than visual examination and radiography. **Recommendation: The characterization plan should clarify under which conditions confirmation of historical AK is warranted and what are the most effective methods proposed.**

Rationale: This finding applies to the approximately the 5 percent of the RH-TRU waste inventory that does not require packaging or repackaging. The committee observes that the March 2002 draft better explains how AK could be, in some cases, the most suitable method to determine the absence of certain prohibited items. In this draft,

DOE proposes to use AK to identify hazardous waste codes to prevent ignitable, corrosive, and reactive waste from being disposed of in WIPP. However, DOE does not clearly explain under what circumstances AK should be confirmed and what are the results expected from supplementary characterization methods. DOE recognizes that visual examination and radiography are inadequate techniques for addressing the entire list of prohibited items.

From the operational experience acquired with CH-TRU waste, it is possible to obtain some indication of the effectiveness of AK as a characterization method. Most of the information collected for CH-TRU waste characterization consists of historical records about waste streams, which the committee termed "historical AK" (see Sidebar 2.1). Effectiveness of AK for CH-TRU waste is demonstrated by the "AK Information Accuracy Reports" from the various generator or storage sites (DOE-CBFO, 2001c). DOE presented to the committee the results of an analysis of the effectiveness of historical AK for CH-TRU waste. Contact-handled TRU waste generators use radiography or visual examination, headspace gas sampling and analysis, and/or solids sampling and analysis to confirm AK information accuracy.

Results showed a high accuracy of historical AK: above 95 percent for the determination of waste matrix codes at the major CH-TRU waste generator sites and above 93 percent for the determination of hazardous waste codes. The only exception was observed at the Idaho Environmental and Engineering Laboratory, where an accuracy of 80 percent was achieved in the hazardous waste codes determination. This lower accuracy was obtained because this site assigned hazardous waste codes on a waste stream basis, rather than on an individual container basis.

Document 2 provides further feedback on AK confirmation for CH-TRU waste by headspace gas analysis:

> "To date, the headspace gas from over 16,000 CH TRU waste containers has been sampled and analyzed and no hazardous waste numbers have been added to a waste stream. This indicates that the additional benefit from using headspace gas sampling and analysis to confirm the hazardous waste determination are limited" (DOE-CBFO, 2002b; pages 5–21).

The committee emphasizes that these effectiveness analyses provided by DOE are relevant only to AK for CH-TRU waste, which is historical AK. In the case of the 95 percent of RH-TRU waste, most of the information comprising AK will be collected during waste generation or repackaging under a certified quality assurance program and would likely have higher accuracy.

Finding 2E: DOE's characterization plan calls for application of specific technologies, such as X-ray radiography, to provide confirmatory data. The committee could not determine the effectiveness of these technologies in characterizing the high-dose-rate fraction of RH-TRU waste containers. **Recommendation: DOE should provide justification for the technologies proposed for obtaining confirmatory data and provide evidence of their effectiveness across the entire spectrum of dose rates for RH-TRU waste.**

Rationale: Given the importance placed on radiography to determine the presence of prohibited items, the characterization plan should clarify and support the information on the method's effectiveness in high radiation fields expected for a small fraction of RH-TRU waste. The lack of information on the effectiveness of characterization methods for RH-TRU waste was also NMED's main argument to prohibit this type of waste in WIPP (see Chapter 3). Moreover, this recommendation is consistent with the Institute for Regulatory Sciences' peer review of the July 2001 draft of the characterization plan.[8]

In Documents 1 and 2, the committee did not find adequate justification for the effectiveness of technologies proposed for obtaining confirmatory data, such as radiography, particularly for RH-TRU waste with higher surface dose rates. In its response to the committee's interim report, DOE states:

> "For the minimal use of NDA [non-destructive assay] and RTR [real-time radiography] that will still be required in the overall program, the standard technologies currently being employed in the CH-TRU waste program are more than adequate. The RH Team recommended program […] calls for NDE [non-destructive examination] to be used only to detect 1 [percent] residual liquid and only for a small percentage of the total RH-TRU waste inventory (most RH-TRU waste will be repackaged or newly generated). Resolution of current RTR systems, even in high radiation fields, is adequate for this determination" (see Appendix H).

DOE never addresses the issue of the fraction of waste containers with high surface dose rates. The current DOE inventory in Table 2.1 shows that both the Hanford Site and Oak Ridge National Laboratory have or will generate RH-TRU waste with a surface dose rate as high as 1,000 rem per hour. Moreover, some Oak Ridge National Laboratory RH-TRU waste also emits neutrons, which further complicates non-destructive characterization methods. In Documents 1 and 2, radiography is cited as lead technology to confirm AK, where necessary, or to examine the contents of small containers prior to placing them into the final shipping containers. DOE's proposal to use radiography across the entire spectrum of dose rates warrants an explanation, however small the fraction of high dose-rate containers may be.

The committee was unable to find elsewhere information on whether radiography could be used for RH-TRU waste containers with the highest dose rates. DOE indicated to the committee that:

> "Though DOE collected some data and analysis indicating that there are no fundamental obstacles to radiographing RH TRU wastes, there has not been a systematic demonstration of that capability. As a consequence, there is lingering doubt regarding its feasibility in general. The simplest means to put these doubts to rest is to design and perform a systematic evaluation" (Taggart, 2001).

The committee gathered additional information on non-destructive characterization techniques and there appear to be uncertainties and technical difficulties in high radiation fields (see Appendix G). Under the committee's request for additional information, DOE provided a publication on X-ray radiography applied to RH-TRU waste characterization (Roney and White, 2001). This study, although performed on surrogate waste, showed that X-ray technology could indeed investigate the content of RH-TRU waste drums, although it was discussed for surface dose rates only up to 100 rem per hour. It is unfortunate that this study is not mentioned in DOE's characterization plan. To

---

[8]The Institute for Regulatory Sciences recommended: "More detail and specificity on WAC [waste acceptance criteria] using AK, VE [visual examination], and [Radiography (including types of instrumentation to be used) should be provided in the permit application" (Institute for Regulatory Sciences, 2001; page 77).

the best of its knowledge, the committee believes that the application of radiography in the presence of very high radiation fields may be problematic.

Other non-destructive assay and examination techniques are being developed for the characterization of RH-TRU waste (see Appendix G). According to the information gathered, the only example of non-destructive assay characterization performed on actual RH-TRU waste was the characterization of 10 canisters of RH-TRU waste at Los Alamos National Laboratory. This was done using a passive/active neutron assay technique (Estep et al., 1989). Further development and application of non-destructive assay and examination techniques is pending the outcome of the final RH-TRU waste characterization plan.

5.3 PROTECTION OF WORKER SAFETY AND HEALTH

The main issues concerning the handling of RH-TRU waste are potential radiation worker doses and assorted characterization costs.

Finding 3A: Potential worker radiation doses and related characterization costs are distinguishing features of RH-TRU waste compared to CH-TRU waste. Available estimates of worker doses and characterization costs for RH-TRU waste are limited, site-specific, and not completely reliable. **Recommendation: DOE could strengthen the rationale of its characterization plan for RH-TRU waste by incorporating a discussion of estimates of worker doses and characterization costs in the three site-specific plans accompanying the submittal documents.**

Rationale: The committee observes that the data presented on waste dose rates and characterization cost estimates in the July 2001 draft have been removed from the March 2002 draft. The committee recognizes that, since the characterization plan for RH-TRU waste is not yet finalized, information on worker doses and characterization costs for RH-TRU waste is scarce. The only data available on RH-TRU waste are from Battelle Columbus Laboratory, WIPP, and Argonne National Laboratory-East, and they may not be representative of all RH-TRU waste generator sites (see Appendix H).

The committee also recognizes that the sites are also responsible for applying the ALARA principle to minimize radiation doses to workers during characterization operations. It follows that each site will produce a different implementation plan for waste characterization and worker radiation protection programs based on ALARA will also be different.

However, a risk perspective of worker doses and costs could be illuminating in developing characterization requirements and in adding credibility to the plan. Discussion of worker doses and costs is relevant since radiation protection standards and criteria are predicated on the concept of assessing the decrement in risk per increment in cost as outlined by the International Commission on Radiological Protection (ICRP, 1977). The approach to estimate worker doses would be to structure different scenarios, rank the scenarios on the basis of the supporting evidence, and calculate the worker doses. This type of dose information would add much to the discussion of the differences between CH- and RH-TRU waste characterization plans. An equivalent approach could be adopted to better estimate characterization costs.

The committee recommends that the estimated doses and costs be an integral part of the three implementation plans from representative generator sites that will accompany Documents 1 and 2. Understanding worker doses and underlying characterization costs is important in developing effective radiation protection programs based on the ALARA principle. The committee provides in Appendix H examples of estimates of worker doses and characterization costs. In Section 5.2, the committee also recommended that DOE's characterization plan should not set forth overly stringent tolerable decision error rates that could negatively impact the sites' ability to implement ALARA (see Finding 2C).

Finding 3B: The submittal documents provide flexibility to the generator sites in the implementation of the RH-TRU waste characterization plan. **Recommendation: DOE should continue its effort in ensuring sufficient flexibility to generator sites in the implementation of the characterization plan. However, characterization activities that share common elements across sites should be standardized.**

Rationale: One of the findings of the committee's interim report is that there is substantial variability among RH-TRU waste generator sites, including:

1. variability in the composition of the waste streams;
2. variability in the extent of AK available;
3. variability in the characterization and repackaging facilities available; and
4. variability and uncertainties in the current and projected inventories of RH-TRU waste.

The committee acknowledges that the March 2002 draft better addresses the substantial RH-TRU waste variability from site to site compared to the earlier draft. DOE's characterization plan allows flexibility to the generator sites in implementing the RH-TRU waste characterization program. In Document 1, DOE directs each TRU waste site to develop standard operating procedures to implement the waste characterization plan at that specific site (DOE-CBFO, 2002a; Section 3.2.2 of Attachment B). DOE also addresses data validation, usability, and reporting at the individual sites in Section 3.5 of Attachment B. Also, Attachment 6 of the application is designated "Site Specific Documentation," but this section has yet to be written. Document 2 requires the sites to develop their own RH-TRU mixed waste characterization programs in compliance with the requirements of the overall waste analysis plan (Document 2, Section R 4.0). These programs are to include characterization strategies, equipment, and health and safety protocols; site-specific characterization methods; quality-assurance plans; and training programs.

Flexibility in a waste characterization plan is important to allow the sites to adapt each characterization requirement to their type of waste inventory and characterization facilities. A rigid, overly prescriptive characterization plan may lead to unnecessary radiation doses to workers and characterization costs. For example, small sites that do not have adequate characterization facilities may find themselves in a difficult situation if the characterization plan mandates specific confirmatory activities, even in the presence of adequate AK. If the site must perform confirmatory measurements or visual examination in a hot cell it would have to ship its waste to a different site[9] equipped with a hot cell or use a mobile hot cell. Therefore, the added costs of such confirmatory measurements could be significant.

The committee reiterates its recommendation that common elements among the sites be standardized to facilitate characterization compliance verifications and, possibly, reduce characterization costs. In the effort of standardizing characterization activities,

---

[9]Shipments from site to site are allowed before the waste characterization plan is approved because DOE's waste characterization plan applies only to waste shipped to WIPP.

DOE will submit to EPA and NMED a sample of completed characterization documents for three RH-TRU generator sites (Energy Technology Engineering Center, Battelle Columbus Laboratories, and Oak Ridge National Laboratory). The committee supports this initiative.

The committee observes that, in the sites' inventories description, DOE did not differentiate mixed RH-TRU waste from non-mixed RH-TRU waste. DOE indicated that, although some waste streams could be identified as non-mixed RH-TRU waste, all waste is deliberately treated as mixed RH-TRU waste, as a policy decision. In fact, according to the RCRA Permit, all waste emplaced in WIPP (mixed or non-mixed) must be analyzed according to a characterization plan approved by NMED (DOE-CBFO, 2002c). The committee acknowledges DOE's policy decision, although it lacks a technical basis.

5.4 COMPLIANCE WITH REGULATORY REQUIREMENTS

The committee evaluated how DOE proposes to address EPA and NMED requirements from a technical point of view. The committee was not asked to determine if the plan complies with the regulatory requirements. The latter is obviously a policy decision that belongs to the regulatory agencies.

One of the challenges of this characterization plan is to meet characterization requirements specified in the EPA Certification and RCRA Permit that were negotiated with WIPP's regulatory agencies for CH-TRU waste and exclude RH-TRU waste. Moreover, the only TRU waste characterization plan approved by EPA and NMED is that currently used for CH-TRU waste. With this characterization plan DOE proposes to meet EPA and NMED requirements following a different approach than that used for CH-TRU waste.

Finding 4A: The proposed characterization plan for RH-TRU waste deliberately tracks the characterization plan for CH-TRU waste. **Recommendation: DOE should evaluate whether existing characterization practices for CH-TRU waste, when applied to the characterization of RH-TRU waste, have an impact on the protection of the environment, health and safety of public and workers, and cost-effectiveness of the characterization program.**

Rationale: DOE proposes to address regulatory requirements for RH-TRU waste characterization using a performance-based approach, a different approach than that used for CH-TRU waste. The committee observes that the characterization plan for RH-TRU waste has not yet been finalized. The implementation activities for RH-TRU waste to address characterization objectives will be finalized only after negotiations based on the submittal documents take place between DOE and WIPP's regulatory agencies. However, the existence of a characterization plan for CH-TRU waste does not imply that the same approach should be proposed for the characterization of RH-TRU waste.

It is not the committee's intention to say that there are unnecessary regulatory requirements in the current EPA Certification and RCRA Permit. However, there is reason to believe (Kehrman, 2002; NRC, 2001 a; DOE, 1999a) that DOE proposed some characterization practices in the CH-TRU waste characterization plan to facilitate negotiations with the regulatory agencies and obtain authorization to operate the WIPP facility. The previous National Research Council committee on WIPP found that some of these self-imposed practices lacked a technical, safety or legal basis (see Sidebar 3.1). The current committee emphasizes that, because of the additional radiological and cost considerations associated with RH-TRU waste, it is even more important that DOE refrains from proposing characterization requirements that do not have a technical, health and safety, or regulatory basis.

Given the different context of RH-TRU waste characterization described earlier, the committee acknowledges that DOE's efforts to introduce a performance-based approach are technically justified. However, the committee does not find the proposed plan to be completely performance based (see Finding 1A). The committee does not suggest that DOE should do only the "bare minimum" to characterize RH-TRU waste. Rather, the characterization program should ensure repository integrity, protection of the environment and of the health and safety of the public and workers, while keeping in consideration tradeoffs between greater characterization accuracy and worker risks.

Finding 4B: The submittal documents include, as basis for characterization objectives, waste acceptance criteria and other requirements than those applicable to the EPA Certification and RCRA Permit. **Recommendation: Submittal documents should focus on regulatory requirements under the relevant agency's purview and should distinguish between these requirements and ancillary information describing the context of RH-TRU waste characterization.**

Rationale: Document 1 contains the RH-TRU Waste Acceptance Requirements and Criteria. These are the "controlling (i.e., the most restrictive) requirements to be used by the sites in preparing their waste for transportation to and disposal at the WIPP" (DOE-CBFO, 2002a; Section 3.0 of Attachment A). The waste acceptance criteria are drawn from transportation and hazardous waste considerations, as well as repository-performance considerations.

For instance, the waste transportation quality assurance requirements are described in Section 4.3 of Attachment A. Since Document 1 is an application for a change in the EPA Certification, it is not clear why requirements and criteria other than those relating to repository performance are given in the document. Historically, DOE included the waste acceptance criteria in the characterization plan for CH-TRU waste (DOE-CAO, 1996a). This does not need to be the case for the RH-TRU waste characterization plan since, to the best of the committee's knowledge, EPA does not require these criteria to be part of the application. Documents 1 and 2 also contain requirements originating from the Nuclear Regulatory Commission, the Department of Transportation, and the Occupational Health and Safety Administration. While it is important to provide the regulatory agencies the full context of RH-TRU waste characterization, including in the submittal documents material outside the regulatory jurisdiction of EPA and NMED may unnecessarily complicate the regulatory review process.

In the future, DOE may wish to negotiate other requirements with different regulatory agencies than EPA and NMED (for instance, transportation requirements with the Nuclear Regulatory Commission and the Department of Transportation). These negotiations could become more complex if the original requirements had also become a part of a document reviewed and approved by the EPA or NMED. See also the comment on polychlorinated biphenyl limits discussed in Finding 2A, example #5.

The committee believes that it would be helpful for both DOE and the regulatory agencies if the submittal documents indicate which requirements are to be reviewed and which are given for information purposes. This would allow the regulator and the implementer (DOE) to keep track of various requirements, and to streamline characterization practices as experience is gained. This recommendation applies to the submittal documents only. The committee supports the current practice of combining all regulatory requirements into one document to guide generator sites in the development of their characterization programs.

5.5 OVERALL ASSESSMENT OF DOE'S CHARACTERIZATION PLAN

The committee observed a net improvement between the July 2001 and March 2002 drafts of the characterization plan. Concerning the plan's technical soundness, DOE itself recognized that this plan is not completely performance based and that several other considerations played a role in the development of this plan. The committee identified some characterization activities lacking technical bases in the context of RH-TRU waste and a potential technical problem with radiographic examination of waste. Also, DOE's proposed characterization plan does not adequately address the issue of tolerable decision error rates associated with all characterization information.

In some instances, the plan lacks specificity because most of the operational details are site-specific and were not available at the time of writing. The site-specific accompanying documents should provide useful clarifications. In the context of RH-TRU waste characterization and from a performance point of view, the committee found that the general approach DOE is proposing is technically sound. However, Documents 1 and 2 do not present a performance-based plan as effectively as they could.

Concerning the plan's protection of worker health and safety, the committee recommends that the approved characterization plan not include overly stringent tolerable decision error rates that could negatively impact the sites' ability to manage worker risks. It is important to recognize that the allowable uncertainties in the final characterization plan approved by EPA and NMED may have an impact on generator sites' radiation protection programs.

Concerning compliance with regulatory requirements in the EPA Certification and RCRA Permit, the committee did not observe any requirement that was not addressed in DOE's characterization plan. In fact, the proposed characterization plan for RH-TRU waste addresses some requirements that are not under the relevant agency's purview. Moreover, the characterization plan for RH-TRU waste deliberately tracks as close as possible that for CH-TRU waste. The committee recommends evaluating whether existing characterization practices for CH-TRU waste, when applied to the characterization of RH-TRU waste, have an impact on the protection of the environment, health and safety of public and workers, and cost-effectiveness of the characterization program.

References

Bhatt, R.N. 2001. Explanation on the origin of RH-TRU waste at INEEL. Electronic communication with National Research Council staff. November 20.

Biedscheid, J., S.Stahl, M.Devarakonda, K.Peters, and J.Eide. 2002. Adequacy of a Small Quantity Site RH-TRU Waste Program in Meeting Proposed WIPP Characterization Objectives. Proceedings of the WM'02 Conference. February 24–28. Tucson, Ariz.

DOE-BCL (U.S. Department of Energy-Battelle Columbus Laboratories). 2002. Electronic communication between Tom Baillieul and National Research Council staff. April 26.

DOE-CAO (U.S. Department of Energy-Carlsbad Area Office). 1995. Remote-Handled Transuranic Waste Study. DOE/CAO-95–1095. Carlsbad, N.Mex. Available at: <http://www.wipp.carlsbad.nm.us/library/cca/cca.htm>.

DOE-CAO. 1996a. Compliance Certification Application, DOE/CAO-1996–2184. Carlsbad, N.Mex.

DOE-CAO. 1996b. Waste Acceptance Criteria for the Waste Isolation Pilot Plant, DOE/ WIPP-069. Revision 5. Carlsbad, N.Mex.

DOE-CAO. 1996c. WIPP Disposal Radionuclide Inventory for the CCA. Revision 3. Table 3–1. June. Transuranic Waste Baseline Inventory Report. Appendix BIR. DOE/CAO-95–1121. Available at: <http://www.wipp.carlsbad.nm.us/library/cca/cca.htm>.

DOE-CAO. 1997. Waste Isolation Pilot Plant Disposal Phase Final Supplemental Environmental Impact Statement. September. DOE/EIS-0026-S-2. Carlsbad, N.Mex.

DOE-CAO. 1998. Transuranic Waste Characterization Quality Assurance Program Plan. CAO-94–1010. Rev. 1.0. December 18. Carlsbad, N.Mex.

DOE-CAO. 1999a. Findings and Recommendations of the Transuranic Waste Characterization Task Force. Final Report. August 9. Carlsbad, N.Mex.

DOE-CAO. 1999b. Waste Acceptance Criteria for the Waste Isolation Pilot Plant. Revision 7. DOE/WIPP-069. Carlsbad, N.Mex.

DOE-CBFO (U.S. Department of Energy-Carlsbad Field Office). 2000. National Transuranic Waste Management Plan. DOE/NTP-96–1204. Revision 2. December. Albuquerque, N.Mex.

DOE-CBFO. 2001 a. Notification of Proposed Change to the EPA's Waste Isolation Pilot Plant 40 CFR Part 194 Certification. July 16. Draft. Revision 1. Carlsbad, N.Mex.

DOE-CBFO. 2001b. Draft Class 3 Permit Modification Request for RH-TRU Mixed Waste. Revision 1. July 17. Carlsbad, N.Mex.

DOE-CBFO. 2001c. Robert Kehrman, DOE-CBFO presentation before the committee during the second information-gathering meeting. October 2. Albuquerque, N.Mex.

DOE-CBFO. 2002a. Notification of Proposed Change to the EPA's Waste Isolation Pilot Plant 40 CFR Part 194 Certification. March. Draft Revision 2. Carlsbad, N.Mex.

DOE-CBFO. 2002b. Draft Class 3 Permit Modification Request for RH-TRU Mixed Waste, Revision 2. March. Carlsbad, N.Mex.

DOE-CBFO. 2002c. Communication between Roger Nelson and National Research Council staff, May 20.

DOE-EM. 2002a. EM Mission and Functions. Office of the Assistant Secretary for Environmental Management. Mission. Available online at: <http://www.em.doe.gov/mission/em01.html>.

REFERENCES

DOE-EM. 2002b. A Review of the Environmental Management Program, United States Department of Energy. Presented to the Assistant Secretary for Environmental Management by the Top-to-Bottom Review Team, February 4. Washington, D.C.

EEG (Environmental Evaluation Group). 1994. Silva, Matthew K; Neill, Robert H. Unresolved Issues for the Disposal of Remote-Handled Transuranic Waste in the Waste Isolation Pilot Plant. Albuquerque, N.Mex.

EPA (U.S. Environmental Protection Agency). 1990. New Mexico: Final Authorization of State Hazardous Waste Management Program. Revision. July 11. Federal Register 55(133):28397.

EPA. 1992. Oak Ridge Federal Facility Compliance Agreement for Mixed Waste Subject to Land Disposal Restrictions (LDR). June 12. Docket No. 92–02 FFR. Available at: <http://www.em.doe.gov/ffaa/orldr.html>.

EPA. 1994. Waste Analysis at Facilities that Generate, Treat, Store, and Dispose of Hazardous Wastes. A Guidance Manual. Office of Solid Waste and Emergency Response (OS-520). OSWER 9938.4–03. April. Washington, D.C.

EPA. 1998. CARD (Compliance Application Review Document) No. 24 Waste Characterization. Available at: <http://www.epa.gov/radiation/wipp/regulations.htm>.

EPA. 2002. Test Methods for Evaluating Solid Waste, Physical/Chemical Methods. SW-846. Third edition. Office of Solid Waste. Available: <http://www.epa.gov/epaoswer/ hazwaste/test/sw846.htm>

Estep, R.J., K.L.Coop, T.M.Deane, and J.E.Lujan. 1989. A Passive-Active Neutron Device for Assaying Remote-Handled Transuranic Waste. LA-UR-89–3736. Paper presented at the Topical Meeting on Non-Destructive Assay of Radioactive Waste. November 17–22. Cadarache, France

Forrester, T.W., R.A.Hunt, and G.L.Riner. 2002. Interim Storage of RH-TRU Waste 72B Canisters at the DOE Oak Ridge Reservation. Proceedings of the WM'02 Conference. February 24–28. Tucson, Ariz.

Gist, C.S., H.L.Plum, C.F.Wu, W.A.Most, T.P.Burrington, and L.R.Spangler. 2002. The Remote-Handled TRU Waste Program Proceedings of the WM'02 Conference. February 24–28. Tucson, Ariz.

Helton, J.C., J.E.Bean, J.W.Berglund, F.J.Davis, K.Economy, J.W.Garner, J.D. Johnson, R.J.MacKinnon, J.Miller, D.G.O'Brien, J.L.Ramsey, J.D.Schreiber, A.Shinta, L.N.Smith, D.M.Stoelzel, C.Stockman, and P.Vaughn. 1998. Uncertainty and Sensitivity Analysis Results Obtained in the 1996 Performance Assessment of the Waste Isolation Pilot Plant. SAND98–0365. Sandia National Laboratories. Albuquerque, N.Mex.

ICRP (The International Commission on Radiological Protection). 1977. Recommendations of the International Commission on Radiological Protection. ICRP Publication 26. Radiation Protection. January 17. Pergamon Press.

Institute for Regulatory Sciences. 2001. Requirements for Disposal of Remote-Handled Transuranic Wastes at the Waste Isolation Pilot Plant. Technical Peer Review Report. ASME International. ASME/CRTD-RP-01–84. July 30-August 3 . Carlsbad, N.Mex.

Kehrman, R.F. 2002. Analysis of Waste Isolation Pilot Plant Waste Characterization Requirements. May 13. Carlsbad, N.Mex.

Knowles, M.K. 2002. Electronic communication with National Research Council staff. July 29.

Knowles, M.K., and K.M. Economy. 2000. Evaluation of Brine Inflow at a Waste Isolation Pilot Plant. Water Environment Research; 72 (4):397–404.

LANL (Los Alamos National Laboratory). 2001. Characterization Parameters, Data Quality Objectives, and Methods for the Remote Handled TRU Waste

REFERENCES

Characterization Program 40 CFR 194 Compliance. Draft. July. Los Alamos, N. Mex.

NMED (New Mexico Environment Department). 1999a. WIPP Hazardous Waste Facility Permit NM4890139088-TSDF. October 27. N.Mex.

NMED. 1999b. New Mexico Environment Department's Direct Testimony Regarding Regulatory Process and Imposed Conditions, RH Waste Prohibition: In the Matter of the Final Permit Issued to the United States Department of Energy and Westinghouse Electric Company Waste Isolation Division for a Hazardous Waste Act Permit for the Waste Isolation Pilot Plant. EPA No. NM4890139088, HRM 98–04(P).

NRC (National Research Council). 1996. The Waste Isolation Pilot Plant: A Potential Solution for the Disposal of Transuranic Waste. National Academy Press. Washington, D.C.

NRC. 2001a. Improving Operations and Long-Term Safety of the Waste Isolation Pilot Plant. National Academy Press. Washington, D.C.

NRC. 2001b. Characterization of Remote-Handled Transuranic Waste for the Waste Isolation Pilot Plant: Interim Report. National Academy Press. Washington, D.C.

ORNL (Oak Ridge National Laboratories). 1989. L.S.Dickerson, Remote-Handled Transuranic Solid Waste Characterization Study: Oak Ridge National Laboratory, ORNL/TM-11050. Oak Ridge, Tenn.

Roney, T.J., and T.A.White. 2001. Characterization of RH-TRU and Lead-Lined Drums Using X-ray Imaging Techniques. INEEL/EXT-2001–00625. July. Idaho Falls: Idaho National Engineering and Environmental Laboratory.

Spangler, L.R., S.M.Djordjevic, R.F.Kehrman, and W.A. Most. 2002. An Alternative to Performing Remote-Handled Transuranic Waste Container Headspace Gas Sampling and Analysis. Proceedings from the WM'02 Conference. February 24–28. Tucson, Ariz.

Taggart, D. 2001. Personal communication with D.Taggart, DOE-CBFO, November 14. The League of Women Voters. 1993. The Nuclear Waste Primer. A Handbook for Citizens by the League of Women Voters. Revised edition. Lyons & Burford, New York, N.Y.

U.S. Congress. 1976. Resource Conservation and Recovery Act of 1976. Public Law 94–580. Available at: <http://www4.law.cornell.edu/uscode/42/6901.html>.

U.S. Congress. 1992. The Waste Isolation Pilot Plant Land Withdrawal Act, as amended. Public Law 102–579. Available at: <http://www.emnrd.state.nm.us/WIPP/lwa.htm>.

Zappe, S. 2001. NMED's Prohibition of Remote-Handled TRU Waste at WIPP. Presentation before the committee during the second information-gathering meeting. October 2. Albuquerque, N.Mex.

Appendix A

Biographical Sketches of Committee Members

Eula Bingham, *Chair,* is professor of environmental health at the University of Cincinnati. Dr. Bingham's interests include risk assessment, regulatory toxicology, environmental carcinogenesis, and occupational health surveillance. Previously, she was a volunteer investigator at the National Institute of Environmental Health Sciences and assistant secretary of labor (Occupational Safety and Health Administration). She is the recipient of the Rockefeller Public Service Award from Princeton University (1980) and the first William Lloyd Award for occupational safety. Throughout her career Dr. Bingham has served on numerous national and international advisory groups, including advisory committees of the National Research Council, Food and Drug Administration, Department of Labor, National Institute for Occupational Safety and Health, National Institutes of Health, Natural Resources Defense Council, and the International Agency for Research on Cancer. These committees addressed research needs in health risk assessment and the potential health effects of environmental exposure to chemicals. In 1989, Dr. Bingham was elected to the Institute of Medicine. She earned her M.S. and Ph.D. in physiology from the University of Cincinnati.

Sanford Cohen is founder and president of S. Cohen & Associates (SC&A, Inc.), an energy and environmental consulting firm providing expertise in radiation sciences, management, health and safety analyses, communications services, and information management. He has managed several contracts for agencies of the U.S. government, including the Environmental Protection Agency, Centers for Disease Control and Prevention, Council on Environmental Quality, Congressional Office of Technology Assessment, Department of Energy, and Nuclear Regulatory Commission. Dr. Cohen is involved in regulatory guidance pertaining to environmental management (including Resources Conservation and Recovery Act/Comprehensive Environmental Response, Compensation, and Liability Act requirements), remediation of contaminated sites, safe disposal of hazardous wastes, site characterization in support of decontamination and decommissioning projects, recycling of scrap metal from nuclear facilities, electric and magnetic fields effects, and indoor radon. Prior to founding SC&A in 1981, Dr. Cohen was vice president and manager of Teknekron, Inc. (Washington Operations), and president of Teknekron Research, Inc., a consulting group working with the above government agencies. Dr. Cohen earned his B.S. in science engineering at Northwestern University and his Ph.D. in nuclear engineering at the University of Michigan.

Milton Levenson is an independent consultant and chemical engineer with more than 50 years of experience in nuclear energy and related fields. His technical experience includes work in technologies related to nuclear safety, fuel cycle, water reactors, advanced reactors, and remote control. His professional experience includes research

APPENDIX A

and operations positions at the Oak Ridge National Laboratory, Argonne National Laboratory, Electric Power Research Institute, and Bechtel. Mr. Levenson is a fellow and past president of the American Nuclear Society, a fellow of the American Institute of Chemical Engineers, and a recipient of the American Institute of Chemical Engineers' Robert E.Wilson Award. He is the author of more than 150 publications and presentations and holds three U.S. patents. Mr. Levenson served as chairman or committee member for several National Research Council studies, including the most recent study on the Waste Isolation Pilot Plant. He was elected to the National Academy of Engineering in 1976. He received his B.Ch.E. from the University of Minnesota.

Kenneth L.Mossman is professor of health physics and director of the Office of Radiation Safety at Arizona State University in Tempe, where he has also served as assistant vice president for research. Prior to his arrival at Arizona State University, Dr. Mossman was a faculty member of the medical and dental schools at Georgetown University in Washington, D.C., and was professor and founding chairman of the Department of Radiation Science of Georgetown's Graduate School. His research interests include radiological health and safety and public policy. Dr. Mossman has authored over 130 publications, including six books and proceedings related to radiation health issues. He has presented testimony before the U.S. House of Representatives and the U.S. Senate. Dr. Mossman served as president of the Health Physics Society and received its prestigious Elda Anderson Award, the Marie Curie Gold Medal, and the Founder's Award. He also served as a Sigma Xi distinguished lecturer. He has been a fellow of the Health Physics Society and the American Association for the Advancement of Science and has served on committees for the National Research Council, National Institutes of Health, Nuclear Regulatory Commission, Nuclear Energy Agency of the Organization for Economic Cooperation and Development (Paris), and the International Atomic Energy Agency (Vienna). Dr. Mossman earned a B.S. in biology from Wayne State University, M.S. and Ph.D. degrees in radiation biology from the University of Tennessee, and an M.Ed, degree in higher education administration from the University of Maryland.

Ernest Nieschmidt is director of the Laser Laboratory and adjunct associate professor of physics at Idaho State University, College of Engineering. His research interests span the development of a neutron activation analysis facility, sonoluminescence, and the destruction of hazardous organic components of mixed waste by free-radical chemistry. He is also involved in the development of the Laser Isotope Separation Laboratory to pursue research into the separation of isotopes using selective excited states induced by laser. For most of his career, Mr. Nieschmidt worked for different contractors at the Idaho National Engineering and Environmental Laboratory on techniques to assay radioactive and transuranic waste material. These included passive and active neutron interrogation, analysis of active gamma rays and gamma rays produced by neutron inelastic scattering, neutron capture, chronoimaging, time correlations, and the application of position-sensitive detectors. He has authored 106 publications related to topics in physical sciences. Mr. Nieschmidt earned his M.S. in physics from San Diego State College.

John Plodinec is director of the Diagnostic Instrumentation and Analysis Laboratory at Mississippi State University. His laboratory specializes in developing instrumentation for unusual environments and in characterizing processes and technologies under real-world conditions. Dr. Plodinec is an internationally recognized expert in waste management and glass science. He has made important contributions in several areas of radioactive waste management, ranging from waste characterization to glass durability modeling. Prior to joining Mississippi State University, he worked for 22 years at the Department of Energy's Savannah River Site, where he collaborated in building and operating the first remote in-cell melter and served as primary technical lead for the product qualification program. In that capacity he oversaw the remote-handled transuranic waste streams produced by the Savannah River Site. He has authored over 90 publications, primarily on waste vitrification and thermodynamics of waste management. He holds a patent on a device to sample high-level waste and a patent on a slurry-feeding device for glass melters. Dr. Plodinec earned his Ph.D. in physical chemistry from the University of Florida.

Anne E.Smith is a vice president at Charles River Associates, an economics consulting firm. Dr. Smith is an expert in risk management, cost-benefit analysis, economic modeling, and integrated assessment of complex environmental and energy issues. Issues that she has analyzed include contaminated sites, global climate change, air quality, and emissions trading. Dr. Smith has developed and reviewed decision support tools for risk-based ranking of contaminated sites and for making risk tradeoffs in selecting remediation alternatives. She has submitted formal comments on the development of the Environmental Protection Agency Hazard Ranking System for identifying Superfund sites, has served on committees of the National Research Council on assessing contaminated site risk management activities, was a project leader in a review for the U.S. Congress of Superfund and Resources Conservation and Recovery Act concerns about the U.S. nuclear weapons facilities, and her testimony has been sought by committees of the U.S. Senate on air quality issues. Dr. Smith has a Ph.D. in economics from Stanford University, with a minor in engineering-economic systems. Prior to joining Charles River Associates, Dr. Smith was a vice president of Decision Focus Inc. She also served as an economist in the Environmental Protection Agency's Office of Policy Planning and Evaluation.

Appendix B

Information-Gathering Meetings

This appendix describes the three committee's information-gathering sessions throughout the study. The committee held an additional meeting devoted to the writing of this report.

B.1 FIRST MEETING: AUGUST 1–2, 2001 (COLUMBUS, OHIO)

Overview of the Battelle Columbus Laboratories: History, Remote-Handled (RH) Waste Inventory, Milestones, Plans for Characterization (James Eide and Craig Jensen, Battelle Columbus Laboratories)

Overview of Characterization Techniques for RH Waste (Greg Becker, Idaho Engineering and Environmental Laboratory)

RH waste at Los Alamos National Laboratory (Larry Field, Los Alamos National Laboratory)

RH Waste at Idaho National Engineering and Environmental Laboratory (Raj Bhatt, Idaho Engineering and Environmental Laboratory)

RH waste at Argonne National Laboratory-West (William Russ, Argonne National Laboratory)

RH Waste at the Hanford Site (Ken Hladek, Waste Management Federal Services of Hanford, Inc.)

RH Waste at Oak Ridge National Laboratory (Gary Riner, Oak Ridge National Laboratory)

Department of Energy (DOE)-Carlsbad Field Office. Introductions. Sponsor's Hopes and Expectations for the Study (Ines Triay, DOE-Carlsbad Field Office)

DOE Strategy for NMED [New Mexico Environment Department] and EPA [Environmental Protection Agency] Submission (Byran Howard, DOE-Carlsbad Field Office)

RH Inventory and Baseline (Joe Harvill, DOE-Carlsbad Field Office)

Repository Performance—RH Impact (Roger Nelson, DOE-Carlsbad Field Office)

Application of Acceptable Knowledge for RH/TRU [Remote-handled transuranic] Waste (Robert Kehrman, DOE-Carlsbad Field Office)

Data Quality Objectives and the State of the Art in Non-Destructive Assay/Non-Destructive Execution Techniques (Dan laggard, Los Alamos National Laboratory)

The RH Program Audit Process—Emphasis on AK (Marlin Horseman, Carlsbad Field Office Technical Assistance Contractor)

Summary: RH-TRU Small—Volume-Small Impact (Ines Triay, DOE-Carlsbad Field Office)

Environmental Protection Agency Overview of the Regulatory Requirements: Criteria to Evaluate DOE's Characterization Plan (Scott Monroe, U.S. Environmental Protection Agency)

Public Comment Period

Field Trip to the Battelle Columbus Laboratories Site-West Jefferson North Campus

B.2 SECOND MEETING: OCTOBER 4, 2001 (ALBUQUERQUE, NEW MEXICO)

Introductions and Opening Remarks (Ines Triay, DOE-Carlsbad Field Office)

Updating the RH Inventory—Recent Information on RH Across the Complex (Joe Harvill, DOE-Carlsbad Field Office)

Doses from RH Waste Disposal: From Generator Site to WIPP Emplacement (Chuan Fu Wu, DOE-Carlsbad Field Office)

Transportation Requirements for RH TRU Waste (Brad Day, DOE-Carlsbad Field Office)

Proposed Standardized RH—TRU Characterization Methods for Non-Destructive Assay (Dan Taggart, Los Alamos National Laboratory)

AK Accuracy Analysis—How Well AK Does for CH Waste Characterization (Sean White and Robert Kehrman, DOE-Carlsbad Field Office)

Results of the Battelle Columbus Laboratories Surveillance—Confirming AK with Supplemental Measurements (Wayne Ledford, DOE-Carlsbad Field Office)

Summary and Open Questions and Answers (Ines Triay, DOE-Carlsbad Field Office)

View of the New Mexico Environment Department (Steve Zappe, New Mexico Environment Department)

RCRA [Resources Conservation and Recovery Act] Requirements According to the Environmental Protection Agency (David Neleigh, U.S. Environmental Protection Agency, New Mexico Office)

Views of the Environmental Evaluation Group (Matthew Silva, Environmental Evaluation Group)
The Other Peer Review: Findings and Conclusion (M.C.Kirkland, Institute for Regulatory Sciences)
Public Comment Period

B.3 THIRD MEETING: FEBRUARY 4–5, 2002 (CARLSBAD, NEW MEXICO)

DOE's Presentations on the New Characterization Plan; DOE's expectations from the Committee; How DOE Plans to Use the Committee's Final Report (various speakers, DOE-Carlsbad Field Office)

New Mexico Environment Department's Comments on the Committee's Interim Report (Steve Zappe, New Mexico Environment Department)

Environmental Evaluation Group's Comments on the Committee's Interim Report (George Anastas, Environmental Evaluation Group)

Environmental Protection Agency's Comments on the Committee's Interim Report (Scott Monroe, U.S. Environmental Protection Agency)

Public Comment Period

Field Trip to the WIPP Site

Appendix C

Excerpt from the Committee's Interim Report: Chapter 5

This appendix consists of an excerpt from the committee's interim report (Box C1). This report focuses on the July 2001 draft of DOE's proposed characterization plan (NRC, 2001 b).

BOX C1 EXCERPT FROM THE COMMITTEE'S INTERIM REPORT: CHAPTER 5

5
Committee's Preliminary Findings and Recommendations

The following are the committee's preliminary findings and recommendations concerning DOE's proposed characterization plan for RH-TRU waste. The committee also provides observations and issues for further clarification to help DOE in future drafts of the plan's supporting documents. As noted previously, findings and recommendations specific to Document 1 and Document 2 are based on the July 2001 draft.

Finding 1: With this new characterization plan for RH-TRU waste, DOE has an opportunity to introduce a truly performance-based characterization plan containing only requirements relevant to the long-term performance of WIPP and that have a safety, technical, or legal basis. **Recommendation: DOE should not include in its characterization plan unnecessary requirements that do not affect the long-term performance of the repository and that do not have a safety, technical, or legal basis.**

Rationale: The committee observes that characterization requirements for RH-TRU waste do not yet formally exist. Characterization requirements specific to RH-TRU waste will be finalized after DOE submits its characterization plan to EPA and NMED. The only legal requirements applying specifically to RH-TRU waste are those in the Land Withdrawal Act (see Chapter 3).

When DOE first applied for the certification of WIPP in 1996, it proposed a characterization plan for both CH- and RH-TRU waste. To facilitate the certification process, DOE imposed extra characterization requirements in the CH-TRU waste characterization plan, which was accepted by EPA and NMED (National Research Council, 2001, Appendix A1). DOE failed to realize that those characterization requirements would become a burden for the generator sites.[1] Also, DOE did not foresee that characterization requirements for CH-TRU waste would become a standard against which to evaluate the RH-TRU characterization plan. In fact, according to the information gathered by the committee, EPA and NMED intend to compare RH-TRU waste characterization requirements to those in the CH-TRU waste characterization plan, as explained in Chapter 4.

The draft RH-TRU waste characterization plan may be following the same path as the CH-TRU waste characterization plan. That is, by wanting to facilitate the certification process for RH-TRU waste, DOE is failing to ask the most important question about the characterization of RH-TRU waste: what is the purpose of waste characterization for WIPP? The answer is the following: to evaluate the impact of waste components on the long-term performance of the repository. The committee recommends that existing requirements for CH-TRU waste, when applied to the characterization of RH-TRU waste, should be evaluated on the basis of their impact on the performance of the repository. The existence of regulatory requirements in the CH-TRU waste characterization plan does not imply that the same requirements should be included in the characterization plan for RH-TRU waste.

Finding 2: The committee questions the relevance of some of the requirements in the RH-TRU waste characterization plan to DOE's stated objective. According to the performance-based evaluation of RH-TRU waste by Sandia National Laboratories, presented in the characterization plan, none of the RH-TRU waste components have an effect on the long-term performance of the repository. **Recommendation: DOE should evaluate characterization requirements in the proposed plan against safety, their impact on the performance of the repository, and regulatory compliance.**

Rationale: The committee questions the relevance of some of the requirements proposed in the characterization of RH-TRU waste. For example, the detection of prohibited items, the determination of metal content, and the attribution of waste summary category groups do not seem to be characterization requirements based on the long-term performance of the repository (see examples below). According to the performance-based evaluation of RH-TRU waste by Sandia National Laboratories (Appendix 1 of Document 1 and Attachment B of Document 2), none of the RH-TRU waste components have an effect on the long-term performance of the repository. Many of the requirements in the proposed characterization plan of RH-TRU waste derive from the CH-TRU waste characterization plan described in WIPP's RCRA permit and EPA's certification. Concerning the CH-TRU characterization plan, the 1998 Committee found that "many requirements and specifications concerning waste characterization and packaging lacked a safety or legal basis"[2] (National Research Council, 2001, page 4). The present committee also recognizes that many such requirements were self-imposed and should be removed from the characterization plan of RH-TRU waste. The committee acknowledges that DOE's approach to use mostly AK without requiring 100 percent confirmatory activities is a positive step in this direction. However, the characterization plan's supporting documents should provide an analysis of the legal and safety requirements proposed for RH-TRU waste on the basis of their impact on the performance of the repository. Of course, legal requirements, like those in the Land Withdrawal Act, cannot be changed or removed, but these legal requirements are few and not excessively restrictive. The committee provides the following examples to support this finding and recommendation:

Example 1: The list of prohibited items contained in the RCRA permit for CH-TRU waste may not be appropriate for RH-TRU waste when the risks associated with hazardous chemical waste are balanced against radiological risks and costs. In Documents 1 and 2, there is no analysis of the safety implications of the prohibited items in WIPP. It is important to analyze the yet-undefined safety implications of these prohibited items and balance these implications against potential radiological risks to workers and associated costs of identifying prohibited items in RH-TRU waste. Such an analysis would be helpful to DOE in comparing the risks workers are exposed to and the associated costs of characterizing RH-TRU waste for prohibited items. The results of the analysis will support and strengthen DOE's RH-TRU waste characterization plan or point out deficiencies in it.

Example 2: In the CCA, DOE wrote: "Ferrous and ferrous-alloy metals (and their corroded products) provide the reactants that reduce radionuclides to lower and less-soluble oxidation states. As discussed in Appendix WCA [Waste Characterization Analysis], the anticipated quantity of these metals to be emplaced in WIPP is two to three orders of magnitude in excess of the quantity required to assure reducing conditions. The waste containers supply more than enough iron to provide adequate reductant. Therefore, no upper or lower limit need be established for the quantity of ferrous and ferrous-alloy metals that may be emplaced, beyond the present projection of containers" (DOE, 1996a, Appendix WCL.2). Nonetheless, in the CCA, DOE imposed a lower limit for ferrous material in the repository at 2×10^7 kg. Nowhere else in the CCA is this limit discussed or justified. This self-imposed limit has somewhat severe consequences because it requires DOE to use steel containers throughout the disposal phase. This example of a self-imposed requirement is acknowledged by DOE-CBFO itself. In fact, DOE informed the committee of its intention to request the elimination of this limit during the re-certification process (DOE-CBFO, 2001f).

Example 3: Waste summary category groups indicate the final form of the waste as it is emplaced in WIPP. These groups are: homogeneous waste, soils and gravel, and debris waste. The CH-TRU waste requirements for NMED direct DOE to attribute a summary category group to each waste container. However, this information does not have any impact on the long-term performance of the repository. According to the information gathered thus far, this information is not used by NMED in any type of decision-making process concerning RH-TRU waste in the repository. Therefore, the committee questions why this requirement is being included in the RH-TRU waste characterization plan. A similar recommendation was made by the Institute for Regulatory Sciences peer review on DOE's characterization plan.[3]

Finding 3: The RH-TRU waste draft characterization plan reviewed by the committee does not clearly present DOE's stated objective of characterizing waste based on its impact on performance of WIPP, while protecting worker safety, reducing costs, eliminating unnecessary self-imposed requirements, and complying with regulatory requirements. In the committee's opinion, the documents reviewed do not put forward a performance-based characterization plan. **Recommendation: In the supporting documents, DOE should clarify the objectives of the characterization plan and how to achieve them.**

Rationale: DOE presented the objectives of its characterization plan for RH-TRU waste during the two information-gathering meetings with the committee. The committee expected these objectives to be clearly presented and pursued in Documents 1 and 2, but this was not the case. Document 2 mentions that DOE's characterization plan adopts a "Performance-Based Measurement System" but the committee struggled to find a clear explanation of this concept and its implementation in the supporting documents. The committee found a clear explanation of a "performance-based characterization plan" in EPA's Compliance Application Review Document No. 24, concerning waste characterization. This document reads: "DOE must provide waste inventory information for use in the performance assessment, including the radionuclide content of waste and the physical and chemical components that may affect disposal system performance" (EPA, 1998, page 21–1). EPA also defines a "waste characteristic" as "a property of the waste that has an impact on the containment of waste in the disposal system" (40 CFR 194.2). DOE's stated objective is in agreement with EPA's statements concerning waste characterization. However, the supporting documents do not explicitly outline the state objectives and the way DOE intends to implement its "performance-based characterization plan." The committee acknowledges that, in Attachment B of Document 1 and in Appendix 1 of Document 2, DOE provides two performance assessment analyses of RH-TRU waste in WIPP. However, these analyses and the performance assessment tool are not fully used to support a performance-based characterization plan. The performance assessment tool played a central role in WIPP's certification process. If the characterization plan were properly built upon performance assessment results, it would be easier for EPA and NMED to assess the plan and it could add credibility to the proposed characterization approach.

Finding 4: From the information gathered during the two committee meetings, it appears that most of the RH-TRU waste to be disposed of at WIPP will be newly generated waste, repackaged waste, or waste that has already been characterized following the CH-TRU waste characterization plan. Therefore, most RH-TRU waste does not need confirmatory measurements because the information collected during repackaging or generation can meet characterization requirements. **Recommendation: In the documents supporting its characterization plan, DOE should discuss the relative volumes of retrievably stored waste and newly generated waste in the context of the different qualities of AK. DOE should also consider the impact of these volumes and AK differences on the characterization plan.**

Rationale: According to DOE, about 80 percent of the RH-TRU waste inventory will be waste to be generated in the future or existing waste that must be repackaged (see Chapter 2). For this

APPENDIX C

large amount of waste, the information constituting AK will be collected during the generation or the repackaging of the waste. The committee agrees with DOE that AK for newly generated waste can meet the data quality objectives developed under the new RH-TRU waste characterization plan. Therefore, for about 80 percent of RH-TRU waste, confirmatory measurements may not be necessary as long as this waste is generated or repackaged using the approved quality assurance program plan.

Documents 1 and 2 do not mention that over 80 percent of the RH-TRU waste inventory will be newly generated or newly packaged waste. The committee recognizes that DOE made the distinction between the two qualities of AK (Attachment A, Document 1) for newly generated waste and retrievably stored waste. However, the committee suggests that the waste characterization plan distinguish not only AK qualities but also the relative volumes of newly generated waste and retrievably stored waste. This information may help DOE, the regulatory agencies, and the public to frame the context of the characterization plan for RH-TRU waste. If the volume of RH-TRU waste represents between 1 and 4 percent of the volume of TRU waste, and 80 percent of RH-TRU waste will meet AK, then only between 0.2 and 0.8 percent of the waste will need confirmatory activities for AK. This is an important observation, and it could have a substantial impact on characterization requirements proposed for RH-TRU waste.

Finding 5: There is substantial variability among RH-TRU waste generator sites, including:

- variability in the composition of the waste streams,
- variability in the extent of AK available,
- variability in the characterization and repackaging facilities available, and
- variability and uncertainties in the current and projected inventories of RH-TRU waste.

DOE's current characterization plan allows some flexibility to the sites, but it does not account explicitly for the above variability. **Recommendation: The RH-TRU waste characterization plan should recognize the large variabilities from site to site and should ensure sufficient flexibility to accommodate them. However, characterization activities that share common elements across sites should be standardized.**[4]

Rationale: The committee received the information about the RH-TRU waste inventories during its two information-gathering meetings. These data are reported in Tables 1 and 2 in Chapter 2. This information is barely present, as supplemental information, in Document 2. in the supporting documents, the committee expected a discussion about RH-TRU waste inventories at the different generator sites, their characterization plans, and capabilities. Including this discussion in Documents 1 and 2 could help DOE to recognize both the large variances from site to site and common elements across sites. Flexibility is important to adapt each characterization requirement to the site's RH-TRU waste inventory and characterization facilities. A rigid, overly prescriptive characterization plan may lead to unnecessary radiation exposures to workers and characterization costs. For example, small sites that do not have adequate characterization facilities may find themselves in a difficult situation if the characterization plan mandates specific confirmatory activities, even in the presence of adequate AK. If the site must perform confirmatory measurements or visual examination in a hot cell it would have to ship its waste to a different site equipped with a hot cell or use a mobile hot cell. Therefore, the added costs of such confirmatory measurements could be significant. The committee also recommends that common elements among the sites be standardized to facilitate characterization compliance verifications and, possibly, reduce characterization costs.

Finding 6: The requirements to qualify information collected on each waste stream, whether by AK or by any other method described in 40 CFR 194.22(b), have not been established with any specificity in the supporting documents. **Recommendation: DOE should revise the supporting documents by adding clear and technically defensible data qualification requirements for its RH-TRU waste characterization plan. Additionally, each data quality objective should have a safety, technical, or legal basis.**

Rationale: To optimize waste characterization activities, it is important to define how generator sites will use AK, radiography, or visual examination to characterize their RH-TRU waste streams, the data qualification requirements, and the procedures to meet those requirements. This recommendation is also consistent with the recommendation of a second peer review on DOE's characterization plan.[6] The committee provides below an example of a missing or an unclear technical basis for one of the data quality objectives in Document 1.

Example: In Document 1, DOE does not provide a technical basis for the requirement that the quantification of the total activity "must be within a factor of 5 of the true value, with a confidence value of 95 percent." DOE provided an explanation during an information-gathering meeting: "The need to accumulate information regarding the total activity of RH-TRU waste is driven by limits set forth in the Land Withdrawal Act. The Land Withdrawal Act limits are derived from original DOE estimates about the RH-TRU inventory. The Transuranic Waste Baseline Inventory Report indicates that the total RH-TRU inventory to be disposed of at WIPP corresponds to a total activity of approximately one million curies (TWBIR, 1996). This corresponds to approximately one-fifth of the Land Withdrawal Act limit (5.1 million curies), hence the factor of five. Allowing an uncertainty of a factor of five (with a confidence level of 95 percent or greater) ensures that, even if the cumulative measurement of the estimated one million curies of RH-TRU waste activity were offset by a factor of five and biased low, the Land Withdrawal Act limit would not be violated (Nelson, 2001). The committee observes that, just because DOE expects to emplace one-fifth of the curies specified in the Land Withdrawal Act, this does not justify a requirement that the measured activity for a unit be within a factor of five of the "true value."

Finding 7: Available estimates of worker exposure and characterization costs for RH-TRU waste are scarce and may not be representative of all RH-TRU waste generator sites. **Recommendation: To better develop and support its characterization plan, DOE should provide more detailed and site-specific estimates of worker exposure and characterization costs for RH-TRU waste. The characterization plan should clearly demonstrate how it minimizes radiation exposure to workers and associated costs.**

Rationale: Since the characterization plan for RH-TRU waste is not yet finalized, information on worker exposure and characterization costs for RH-TRU waste is scarce. The only data available are from BCL and LANL, and they may not be representative for all RH-TRU waste generator sites (see Chapter 2). However, exposure and cost estimates can be calculated quite well, particularly if the uncertainties become a part of the results. The committee found the risk/cost impact study in Document 2 (Supplemental Information) to be very informative. However, this study only considered three characterization scenarios to estimate exposure risks and costs: the full characterization used for CH-TRU waste (see Appendix A), AK only, and visual or radiographic examination. A detailed risk perspective of worker exposure and costs could be very illuminating in developing characterization requirements and in adding credibility to the plan. The approach to estimate exposures would be to structure different scenarios, rank the scenarios on the basis of the supporting evidence, and calculate the exposures and doses. This type of dose and exposure information would add much to the discussion of the differences between CH-and RH-TRU waste characterization plans. An equivalent approach could be adopted to better estimate characterization costs. A similar recommendation was also provided by Restrepo and Millard.[7]

Finding 8: DOE's characterization plan calls for application of specific technologies, such as X-ray radiography, to provide confirmatory data. The committee could not determine the effectiveness of these technologies in characterizing the high-dose-rate fraction of RH-TRU waste containers. **Recommendation: DOE should provide complete and defensible justification for the technologies proposed for obtaining confirmatory data and provide evidence of their effectiveness.**

Rationale: DOE specifically mentions "X-ray scanning of waste containers to identify and verify waste contents" (Document 1, Attachment A, page 15). Given the importance placed on radiography to determine the presence of prohibited items, the characterization plan should clarify and support the information on the method's effectiveness in high radiation fields expected for RH-TRU waste. In general, DOE should provide complete and defensible justification for the technologies proposed for obtaining confirmatory data and provide evidence of their effectiveness. This issue was also raised by NMED during the committee's information-gathering meeting (Zappe, 2001). This recommendation is also consistent with a recommendation by the RSI peer review on DOE's characterization plan.[8] DOE also concurs on this matter by stating: "Though DOE collected some data and analysis indicating that there are no fundamental obstacles to radiographing RH-TRU wastes, there has not been a systematic demonstration of that capability. As a consequence, there is lingering doubt regarding its feasibility in general. The simplest means to put these doubts to rest is to design and perform a systematic evaluation" (Taggart, 2001 e). The committee supports this suggestion.

Observations and Issues for Further Clarification

To improve understanding, corroborate, and add credibility to its characterization plan, DOE should clarify the following points in its supporting documents:

Observation 1: The committee observes that the information in the supporting documents for the RH-TRU waste characterization plan is sometimes convoluted, difficult to understand, difficult to find, and repetitious. Examples are provided below.

Example 1: DOE presents an analysis that shows that the gamma-emitting radionuclides in RH-TRU waste will have a minimal impact on the long-term performance of the repository. DOE also presents an analysis that shows that RH-TRU waste components (metals, free liquids, cellulosics, plastic, rubber, and hazardous material) will have a minimal impact on the repository performance. These analyses are presented in Attachment B of Document 1 and in Appendix A of Document 2, respectively. The body of the text supporting the characterization plan barely refers to these two key analyses. DOE should give these analyses a more prominent role in Documents 1 and 2 and explicitly refer to these analyses to support its characterization plan.

Example 2: Some aspects of the characterization plan are difficult to evaluate because terms such as "representative selections" and "appropriate samples" are not properly defined.

Example 3: The plan, as described in Documents 1 and 2, lacks specificity. For example, for the CPR and free water, when is AK and when is radiography employed? Are surveys conducted by the same mechanisms for lower and higher surface dose rates? What is the link between the surface dose rates and the total curies by waste stream? How is a waste stream determined to be suitable for disposal at WIPP (i.e., transuranic waste of defense origin)? How are the ten radionuclides important to the long-term performance of WIPP (see Sidebar 2) identified and quantified?

Observation 2: There are conflicting statements and discrepancies between Document 1 and Document 2. The committee acknowledges that the two documents address different agencies and requirements. However, to improve the understanding and add credibility to its characterization plan, the documents could be made consistent with each other and conflicts and discrepancies resolved whenever possible. The committee provides three examples of conflicts or discrepancies between Document 1 and Document 2.

Example 1: Document 1 states that the differences between the RH- and CH-TRU characterization plans are not significant: "Because the RH-TRU waste characterization program does not represent a significant change to the existing certification, the 40 CFR 194 process will be the compliance mechanism for obtaining EPA's approval of each individual TRU waste site for disposal of its RH-TRU waste at WIPP" (Document 1, EPA01 Notice, page 6). However, Document 2 finds the differences significant: "The proposed modification is classified as a Class 3 permit modification per 20.4.1.900 NMAC…As indicated above, the additional provisions being requested in this permit modification, deemed necessary to provide for management, storage, and disposal of RH-TRU wastes, are indeed substantial and therefore meet the criteria and intent of a Class 3 modification" (DOE, 2001b).

Example 2: In Document 2, DOE allows for estimates of CPR, metals, and volatile organic compounds in the characterization of these waste constituents. Yet, DOE requires measurements without any specification of allowable uncertainties of the quantities of free water. This appears to be in conflict with the analysis in Document 1, Appendix A, which concludes that the performance of the repository is insensitive to the water content of the waste.

Example 3: Some of the statements made in Document 2 appear to contradict the approach to waste characterization discussed in Document 1. For example, on page 2–5 of ITEM 2, Document 2, DOE states: "This demonstrates that the waste material parameter weights and amount of free liquid for RH-TRU waste are not needed to ensure the integrity of the repository; therefore, waste characterization activities to identify these parameters are not necessary." Yet in Document 1, DOE lays out a program to measure the amount of free liquid in each waste container. The supporting documents should clearly specify that free liquid limits are necessary to meet transportation requirements and are not directly relevant to repository performance.

Observation 3: It is not clear how visual examination and radiography can differentiate all prohibited from non-prohibited items. For example, visual examination and radiography cannot distinguish between corrosive and non-corrosive liquids, whereas AK may provide records of the existence of such liquids in the waste. Therefore, AK may be a better indicator of some of the currently prohibited items than visual examination and radiography.

Closing Remarks

The statement of task directs the committee to provide recommendations based on scientific and technical considerations. The committee recognizes that some of its recommendations may not be implemented because of regulatory or policy decisions, which may not be based on scientific or safety considerations. These decisions ultimately belong to the regulatory agencies. The committee acknowledges and supports DOE's endeavors to improve worker safety, reduce costs, and eliminate unnecessary self-imposed requirements. However, the documents prepared for EPA and NMED to present the characterization plan for RH-TRU waste do not address these goals as effectively as they could. The committee acknowledges that DOE has been extremely responsive in revising the characterization plan on the basis of discussions during the information-gathering meetings. Also, staff of DOE-CBFO efficiently provided additional information whenever requested. The committee looks forward to future drafts of the characterization plan for RH-TRU waste. The committee will hold two more meetings to discuss the next draft of the characterization plan (if available), further address issues identified in this interim report, and develop its final report, which will be issued in the summer 2002.

---

[1] DOE, during the second committee meeting, declared that the "contact-handled TRU waste characterization program is inefficient and costly" (DOE-CBFO, 2001d).

[2] According to the 1998 Committee, the requirements that do not have a technical basis are: sampling and analysis of homogeneous waste, headspace gas sampling and analysis, and visual examination procedures to characterize CH-TRU waste (see Chapter 4).

[3] The RSI peer review recommended: "DOE should evaluate the necessity of identifying waste streams by EPA's Hazardous Waste Numbers or Characteristics. If there is no impact on WIPP performance and integrity, the DOE should work with the regulatory agencies to remove this requirement" (RSI, 2001, page 77).

[4] According to the information provided by DOE, DOE is already amending its characterization plan to standardize as much as possible common elements across the complex.

[5] Shipments from site to site are allowed before the waste characterization plan is approved because DOE's waste characterization plan applies only to waste shipped to WIPP.

[6] The RSI recommended the following: "A detailed procedure for determining whether there is sufficient AK available on a waste should be developed as part of the permit application. This procedure should be consistent across all waste generating sites…[A] detailed procedure should be provided to go to other characterization methods if AK is found to be insufficient" (RSI, 2001, page 77).

APPENDIX C

[7] The authors of the risk/cost analysis study in Document 2 recommend that "a detailed risk/cost impact analysis should be performed for individual site generators to quantify waste characterization option costs, potential worker dose rates, worker risks, and overall impact. This would facilitate a more rigorous evaluation of risks and costs for characterization options" (Restrepo and Millard, 2001, page iii).

[8] In its peer review, the RSI recommended: "More detail and specificity on WAC [waste acceptance criteria] using AK, VE [visual examination], and radiography (including types of instrumentation to be used) should be provided in the permit application" (RSI, 2001, page 77).

NOTE: References in this appendix apply to the committee's interim report (NRC, 2001b).

Appendix D

DOE's Response to the Committee's Interim Report

In response to the committee's interim report, the Department of Energy (DOE)-Carlsbad Field Office transmitted a summary of proposed resolutions or action items for the following draft of its characterization plan.

D.1 DOE'S RESPONSE

DOE's response to the committee's interim report consists of two following two elements: 1) a letter containing DOE's proposed resolution and action items as a result of the committee's findings and recommendations and 2) an attachment to the letter, named Attachment 1. The letter and its attachment are presented in Boxes D.1 and D.2, respectively.

BOX D.1 DOE'S RESPONSE TO THE COMMITTEE'S INTERIM REPORT

PROPOSED RESOLUTION/ACTION ITEMS

Concerning NAS Finding and recommendation #1: We have reevaluated the characterization information needs consistent with this recommendation. The process considered long-term repository performance information needs, safety and technical considerations, regulatory requirements, and legal requirements and led to the development of a discreet list of Characterization objectives. Attachment 1 is the result of this effort. Attachment 1 [see below] is being used to revise the Characterization Implementation Plan in the EPA submittal, as well as the WAP in the RCRA submittal. Attachment 1 also indicates whether or not we believe the characterization objectives require confirmation, and the methods we will propose in the submittal documents to obtain the characterization information.

Concerning NAS Finding and recommendation #2: The aforementioned Attachment 1 is the result of the reevaluation process we conducted as proposed by the NAS WIPP Panel.

Example 1: The list of prohibited items in the draft WAP (July 2001) will not be included as a specific list in the next draft WAP. With the exception of residual liquids, these "items" will be excluded from the list of prohibited items, but will continue to be dealt with by determining the hazardous waste numbers and prohibiting D001, D002, and D003. This effectively drives the decision from the container level to the stream level where hazardous waste code assignment decisions are typically made. The hazardous waste code(s) determination will be accomplished using AK. The 1% free liquid criteria was retained as a RCRA operational compliance consideration rather than for safety or performance reasons. The next draft WAP will be modified consistent with this change. See Attachment 1 for the rationale and justification for hazardous waste numbers and residual liquids.

Example 2: CBFO may, in the future, choose to remove the metals limits set in the CCA.

APPENDIX D

Until this change is made, the proposed RH-TRU characterization program must continue to require tracking of metals. Contrary to the NAS statement ("severe consequences") the CBFO believes the "limits" do not impact the program, and the proposed characterization method of counting containers is simple to implement. For other reasons steel is the preferred material for waste containers, but the "limit" does not prohibit the use of other materials. The DOE anticipates that a minor amount of CH-TRU waste and no RH-TRU waste will be shipped in non-metal containers. We therefore do not view the consequences of the limitation as severe.

Example 3: The NAS is correct in pointing out that Summary Category Groups (SCG) have no impact on long-term repository performance. SCG is in the current CH RCRA Permit in order to affect a decision on what kind of waste analysis is required by the Permit (S3000 and S4000 waste streams must be sampled and chemically analyzed). No sampling and analysis is proposed for RH-TRU waste, however the SCG requirement is retained in the program because it allows for an easy method of accounting for cellulosics, plastics, and rubber (CPR), which are related to repository performance. Since SCG determination is relatively easy and is part of the EPA program anyway, it will be included in the RCRA program as an activity complying with 40CFR264.13.

Concerning NAS Finding and recommendation #3: The re-evaluation discussed for Finding #1 considered characterization objectives from a performance driven perspective as well as from regulatory and legal aspects. *Nearly all resulting characterization objectives are unrelated to repository performance or to safety/technical considerations.*

New supporting documentation will clearly show the basis for each characterization objective.

Concerning NAS Finding and recommendation #4: Consistent with this recommendation, DOE is preparing completed characterization documents for three major RH-TRU sites. These documents will be submitted directly to the EPA for its approval through the formal rulemaking process, and they will also be submitted to the NMED as examples along with a new draft request for RCRA Permit modification.

Concerning NAS Finding and recommendation #5: Submittals to both the NMED (RCRA compliance) and the EPA (40CFR194 compliance) are being redrafted to balance flexibility with standardization as appropriate.

The characterization documents discussed for Finding #4 will provide detailed inventory information which will assist in determining what aspects of the program can be standardized and how to allow for flexibility in site programs.

Concerning NAS Finding and recommendation #6: This Finding and Recommendation is also fully addressed by the Resolution discussed for Finding #1.

DOE agrees with the Panel that the former DQO associated with Total Activity is legally based and not technically based. As a result of the re-evaluation process described in the Resolution to Finding #1, it will be removed from the draft regulatory submittal documents.

Concerning NAS Finding and recommendation #7: After re-evaluating the approach to RH-TRU waste characterization, as discussed in Finding #1, DOE continues to believe that nearly all required RH-TRU waste characterization can be accomplished using AK. Therefore worker exposure is no longer a significant issue related to RH-TRU waste, and eliminating most characterization physical activities has minimized program costs. Characterization activities may still be required on a very small percentage of the RH-TRU waste inventory, but these activities will always be conducted in such a way as to minimize worker exposure.

Concerning NAS Finding and recommendation #8

As suggested by the Panel, DOE will complete the "systematic evaluation" recommended by Taggart to conclusively demonstrate that current RTR technology is adequate for RH-TRU waste characterization use; however, the revised characterization program no longer requires that confirmatory data be generated. For the minimal use of NDA and RTR that will still be required in the overall program, the standard technologies currently being employed in the CH-TRU waste program are more than adequate.

The RH Team recommended program (see Attachment 1) calls for NDE to be used *only to detect 1% residual liquid and only for a small percentage of the total RH-TRU waste inventory* (most RH-TRU waste will be repackaged or newly generated). Resolution of current RTR systems, even in high radiation fields, is adequate for this determination.

Concerning NAS Observation #1: Example 1: The next draft of the regulatory submittals will clearly reference the supporting analyses, and each characterization objective will have an explicit pointer to its basis.

Example 2: The redrafted documents will be purged of imprecise language and important terms will be properly defined.

Example 3: After the re-evaluation discussed in Finding #1, CPR will not be measured, it will be conservatively assigned a weight (based on the analyses referred to in Observation #1). However, in general, other techniques are to be employed only for the few (and usually small) waste streams for which there are insufficient AK. There is no direct link between surface dose rate and total curies that is used by the proposed program, however in some individual cases for some individual waste streams, if enough is known about the waste stream (AK), surface dose rate measurements can be used to adequately estimate the activity. The Defense Waste determination for RH-TRU waste will follow the acceptable precedent that has been established in the CH-TRU waste characterization program. The modified EPA submittal will provide detail concerning the TRU versus LLW determination. The ten radionuclides *are not important to repository performance* (See RH Impact Assessment included in the EPA submittal). This requirement will therefore not be included in the revised submittal document.

Concerning NAS Observation #2: Example 1: The statement made in Document 1 (the EPA submittal) was made in the context of what is significant relative to 40CFR194 and also allows that the EPA has the final decision regarding significance. The statement made in Document 2 (the RCRA submittal) was made in the context of what is significant relative to 20.4.1.900 NMAC. In addition, the NMED has made it very clear that they consider this to be a Class 3 modification.

Example 2: The modified Documents will no longer require estimates for CPR, metals, or VOCs. Some method of assurance that there is less than 1% residual liquids in a waste container is still required to address RCRA operational requirements, however no statement of allowable uncertainties is derived since there is no technical basis for an uncertainty statement.

Example 3: The redrafted supporting documents will make it clear that the origin of the 1% liquid requirement is from transportation, has no bearing on repository performance, and is only indirectly related to RCRA regulatory requirements.

Concerning NAS Observation #3: *The NAS is correct in the observation that VE and RTR are inadequate techniques for distinguishing the list of prohibited items in the previous draft.* RTR or VE are suitable techniques for detecting and distinguishing residual liquids, and the reevaluation requested in Finding #1 has determined that RTR or VE will only be used for that purpose. See also the Resolution to Finding #8.

NOTE: This table has been edited for formatting purposes. The emphasis (italicized and underlined text) was added by DOE-Carlsbad Field Office.

SOURCE: DOE (DOE-CBFO, 2002).

APPENDIX D

> **BOX D.2 ATTACHMENT 1 ACCOMPANYING DOE'S LETTER TO THE COMMITTEE (SEE BOX D.1)**
>
> **ATTACHMENT 1**
> **RH-TRU WASTE CHARACTERIZATION PROGRAM**
> **Objectives, Characterization Methods, Implementation Standards RH-TRU Team Rationale and Justification**
> **Introduction**
>
> The RH-TRU Team has reached consensus on what the basic elements of an RH-TRU waste characterization program should consist of (i.e., characterization objectives, characterization method, and implementation). This consensus was achieved by taking a "performance based approach" and examining from the "ground up" what the minimum characterization requirements should be to satisfy 40CFR194.24 (EPA) requirements and RCRA (NMED) requirements. Previous recommendations, findings, and observations from RSI and NAS/NRC reviews were heavily considered, and as the re-evaluation process progressed, each potential requirement was examined with respect to its basis; that is safety, repository performance, legal, or regulatory.
>
> Table 1 for EPA requirements and Table 2 for NMED requirements present the RH-TRU Team consensus recommendations of what basic characterization elements should be used to build an EPA and NMED compliant RH-TRU waste characterization program. This paper explains the RH-TRU Team rationale or justification for each entry in the attached Tables, [see Box D.1]
>
> **Table Definitions (Columns) and Entry Justification/Rationale**
>
> There is a separate Table for the EPA related RH-TRU waste characterization elements and the NMED related RH-TRU waste characterization elements, however the column definitions are identical for both.
>
> <u>Characterization Objectives</u>: This column lists the necessary overall program objective that must be accomplished by collecting waste characterization information. It is the basic question being asked and stems from safety, repository performance, regulatory, or legal considerations.
>
> **EPA Characterization Objectives**
>
> 1. *Account for ferrous and non-ferrous metals,* is related to repository performance in that the CCA Appendix WCL requires a minimum of metals (20 million kilograms of ferrous metals and 2,000 kilograms of non-ferrous metals) be included in the final WIPP repository inventory. Thus it is necessary to account for disposed metals, <u>by some means</u>, in order to ensure that the specified minimums are emplaced. Once the minimum is achieved, no further accounting is necessary.
> 2. *Account for Cellulosics, Plastics, Rubber (CPR),* is related to repository performance in that the CCA Appendix WCL limits total WIPP repository disposed inventory of CPR to a maximum of 20 million kilograms. Thus it is necessary to account for disposed CPR inventory, <u>by some means</u>, in order to ensure that the 20 million kilograms is not exceeded.
> 3. *Account for Free Water,* is a regulatory expectation that the RH-TRU Team believes is prudent to meet even though no amount of free water in the RH-TRU waste will impact repository performance. Note that CCA Appendix WCL.5 states that, "Consequently, there is no need to monitor water in the waste for compliance with 40 CFR § 194.24(c)." The RH-TRU Team recommends that free water in RH-TRU waste be accounted for <u>by some means</u>.

4. *Account for TRU Activity,* is a regulatory expectation that the RH-TRU Team believes is prudent to meet even though, for RH-TRU waste, it is unrelated to repository performance. Note that CCA Appendix WCL.1 states that, "The total activity of the waste is not important for 40CFR194.24(c), because the containment requirements are normalized to the initial inventory." The relationship of TRU activity to total activity is also explained in WCL.1, "The total activity of the waste at emplacement and during the entire 10,000-year performance period is dominated by the activities of four emplaced radionuclides: Am241, Pu238, Pu239, and Pu240." The RH-TRU Team recommends that TRU activity in RH-TRU waste be accounted for <u>by some means</u>.

5. *Ensure waste is TRU (activity`100 nCi/gm),* is a legal requirement from the WIPP Land Withdrawal Act (LWA). The Team is in agreement that it is prudent to include it in the EPA Table, even though some members of the Team do not believe it to be of EPA regulatory concern.

6. *Account for Total Activity,* is related to the LWA legal requirement that WIPP final RH-TRU waste inventory be limited to 5.1 million total curies. Thus it is necessary to account for disposed RH-TRU waste total activity, <u>by some means</u>.

7. *Limit canister activity to <23 curies per liter,* is a LWA invoked legal requirement. This limit is specifically included in 40CFR194 as an EPA regulated Characterization Objective.

8. *Limit Surface Dose Rate of each container to <1000 rem/h,* is a LWA invoked legal requirement. This limit is specifically included in 40CFR194 as an EPA regulated Characterization Objective.

9. *Limit WIPP inventory to <5% by volume for >100 rem/h on a canister basis,* is a LWA invoked legal requirement. Thus canisters with surface dose rate exceeding 100 rem/h must be identified <u>by some means</u>.

10. *Ensure waste is RH (surface dose `200 mrem/h),* is a de facto legal requirement of the LWA. The Team is in agreement that it is prudent to include it in the EPA Table, even though some members of the Team do not believe it to be of EPA regulatory concern.

NMED Characterization Objectives

1. *Assign Hazardous Waste Numbers,* is related to the regulatory requirement to comply with the terms and conditions of the WIPP Hazardous Waste Facility Permit (HWFP). The WIPP HWFP specifically prohibits hazardous wastes with the numbers D001, D002, and D003. Some other hazardous waste numbers have not been applied for and therefore are not allowed by the HWFP.

2. *Identify Physical Form,* is a regulatory requirement that is indirectly related to 40CFR264.13.

3. *Limit residual liquids to <1% volume of RH canister (or drum for 160B),* is related to a regulatory requirement to provide secondary containment at the WIPP for potential spills. An analysis justifying the 1% for CH-TRU waste at the WIPP is part of the HWFP record, however a similar analysis has not been done for RH-TRU waste. The RH-TRU Team believes it is prudent to include this limit as an RH-TRU waste requirement also.

Note 1: The current HWFP for CH-TRU waste also includes residual liquid restrictions for internal containers (i.e., "Waste shall contain as little residual liquid as is reasonably achievable by pouring, pumping and/or aspirat-ing, and internal containers shall contain less than 1 inch or 2.5 centimeters of liquid in the bottom of the container."). The RH-TRU Team did not retain these requirements for RH-TRU waste since their main purpose was related to eliminating the need to characterize liquids in internal containers (i.e., containers are RCRA "empty"). NMED Characterization Objective Number 1, *Assign Hazardous Waste Numbers,* requires the assignment of appropriate hazardous waste codes to containers, and the driver to further characterize liquids is eliminated.

Note 2: In addition to liquids the September 2001 draft RH-TRU waste WAP included the following "prohibited items:"

- Pyrophoric materials
- Incompatible waste
- Explosives and compressed gases
- PCBs with concentrations greater-than-or-equal-to 50 ppm

The RH-TRU Team concluded that it is unnecessary to specifically identify these "items" in the WAP or as specific Characterization Objectives. Pyrophoric materials, explosives and compressed gases fall under the RCRA definition of reactives (D003) and therefore are precluded by NMED Characterization Objective number 1, *Assign Hazardous Waste Numbers*. All wastes with HWFP acceptable Hazardous Waste Numbers have been demonstrated to be compatible, and therefore NMED Characterization Objective number 1, *Assign Hazardous Waste Numbers,* assures that there are no incompatible wastes. PCBs are subject to EPA regulation under TSCA and are regulated regardless of whether or not they are included in a RCRA permit as a prohibition. Therefore the RH-TRU Team has not included any of these "prohibited items" in the proposed RH-TRU waste characterization program as specific characterization objectives.

4. *Identify and Quantify VOCs,* is related to health and safety considerations derived from other (than RCRA) EPA release and exposure limits imposed by the NMED. The existing VOC conditions in the HWFP (e.g. Room Based Limits for VOCs) are based on an extremely conservative analysis of bounding VOC releases. The RH-TRU Team recommends that a bounding analysis specific to the small volume of RH-TRU waste will obviate the need for generator/storage sites to characterize RH-TRU waste for VOCs. This RH-TRU Team recommended approach for VOCs is exactly that already proposed in the September 2001 draft Class 3 Request for RCRA Permit Modification for RH-TRU waste.

Characterization Method: This column lists the general method to be used (as recommended by the RH-TRU Team) to accomplish the *Characterization Objectives* from the first column.

EPA Characterization Methods

1. *Count containers emplaced in the WIPP,* is a simple administrative accounting using the WIPP Waste Information System (WWIS) that meets the metals Characterization Objective. This method has already been approved by the EPA in the 40CFR191/194 "Final Rule" and is currently employed in the CH-TRU waste program. This is not a generator/storage site characterization requirement; it is a WIPP administrative requirement.

2. *Characterization method—Use AK to determine Summary Category Group (SCG) of each waste stream. WIPP admin requirement—Use WWIS to assign 620 kg/m^3 up to container tare weight for debris waste.* Relative to the CCA total inventory limit of 20 million kilograms of CPR, any practical contribution to the CPR inventory from RH-TRU waste is inconsequential. The *RH-TRU Inventory Impact Assessment Report* (Sandia, June 2001) used 620 kg/m^3 of plastic for a bounding analysis that demonstrated

repository performance is not affected by RH-TRU waste CPR. Out of the three Summary Category Groups, (S3000—homogeneous solids, S4000—soils and gravel, S5000—debris) only S5000—debris contains significant quantities of CPR. Thus a very simple and conservative approach for meeting the CPR Characterization Objective is to use AK to identify S5000—debris waste streams, and use the WWIS to electronically calculate and assign a CPR weight value for every S5000—debris RH-TRU waste container that is emplaced in the repository. <u>Because this assignment is based on a bounding analysis, and because there will always be enough AK to make the simple SCG designation, no confirmation of this designation is required</u>.

3. *Use AK to assign either 0% or 1% by volume for each waste stream.* The *RH-TRU Inventory Impact Assessment Report* (Sandia, June 2001) assumed that the final RH-TRU waste inventory in the WIPP was 50% by volume water. This bounding analysis demonstrated repository performance is not affected by any amount of water in RH-TRU waste. Thus a very simple and conservative approach for meeting this Characterization Objective is to use AK to either assign 0% volume of liquid to a specific RH-TRU waste stream or assign 1% volume. The RH-TRU Team chose 1% rather than 50% because transportation requirements will not allow more than 1% to be transported. <u>Because the Sandia bounding analysis effectively demonstrates that selecting either 0% or 1% makes no difference in repository performance, and because AK will, in some cases, allow a definitive 0% decision to be made for a waste stream (e.g., thermally treated waste) or a default 1% will be assigned, no confirmation of this AK determination is required</u>.

4. *Use AK to determine the relationship of TRU Activity to Total Activity for each waste stream, or establish a waste stream value by sampling and measurement, or measure the TRU Activity of each container.* The *RH-TRU Inventory Impact Assessment Report* (Sandia, June 2001) assumed, for a bounding analysis, that the entire RH-TRU waste curie loading allowed by the LWA (5.1 million curies) was composed of Pu239 and showed that even this impossible condition would still comply with the 40CFR191 containment requirement. Thus any minimal AK information available with which to roughly estimate the ratio of TRU to Total Activity is sufficient to meet this Characterization Objective <u>without the need for confirmation</u>. If this minimal AK information is not available, measurements of Total Activity and TRU Activity should be obtained on a few randomly selected containers from the waste stream to establish a waste stream value. Alternatively a generator/storage site may choose to measure the TRU Activity of each container. Note that the method for meeting this Characterization Objective (EPA number 4) is closely tied to, and interrelated with, the following two methods for EPA Characterization Objectives numbers 5 and 6.

5. The same Characterization Method used for TRU Activity (Characterization Method No. 4) is also used to make the TRU determination. Usually AK can be used to make a determination of TRU (versus LLW) for an entire waste stream if there is loading information (activity or mass) regarding Pu238 or Am241 and Pu239. Since this is a "legal" limit with no safety or repository performance implications, <u>this AK determination need not be confirmed</u>. If the TRU determination cannot be made using AK, a waste stream TRU determination may be made by measuring a few randomly selected containers. A TRU determination may also be made by measuring the value of TRU concentration for each container.

6. *Use AK to determine Total Activity of the waste stream, or establish a waste stream value by sampling and measurement, or measure the Total Activity of each container.* Any minimal AK information available with which to roughly estimate the Total Activity is sufficient to meet this Characterization Objective <u>without the need for confirmation</u>. *Characterization Parameters, Data Quality Objectives, and Methods* (LANL, July 2001) pointed out that the expected (from the TWBIR) RH-TRU waste total activity inventory could be biased low by a factor of five before the LWA limit is challenged. If this minimal

APPENDIX D

AK information is not available, measurements of Total Activity should be obtained on a few randomly selected containers from the waste stream to establish a waste stream value. Alternatively a generator/storage site may choose to measure the Total Activity of each container.

7. The same Characterization Method used for Total Activity (Characterization Method No. 6) may also be used to meet Characterization Objective number 7. *Characterization Parameters, Data Quality Objectives, and Methods* (LANL, July 2001) showed that for the vast majority of RH-TRU waste streams AK will demonstrate that it is not feasible to load a canister (or any container) to 23 curies per liter. For the rare exception where AK indicates 23 curies per liter is hypothetically feasible, measurements of Total Activity should be obtained on a few randomly selected containers from the waste stream to establish a waste stream value. Alternatively a generator/storage site may choose to measure the Total Activity of each container.

8. *Use standard industry survey methods to measure and report Surface Dose Rate.* This is a standard practice measurement using standard equipment. Any container that has a measured Surface Dose Rate of 1000 rem/h or greater cannot be shipped to the WIPP. This measurement also is used for achieving Characterization Objectives 9 and 10.

9. *Use standard industry survey methods to measure and report Surface Dose Rate.* This is a standard practice measurement using standard equipment. The Surface Dose Rate measured value of each container that is shipped to the WIPP must be reported in the WWIS. The WIPP will use WWIS data to load manage and assure no more than 5% of the containers of RH-TRU waste emplaced in the repository exceed a Surface Dose Rate of 100 rem/h.

10. *Use standard industry survey methods to measure and report Surface Dose Rate.* This is a standard practice measurement using standard equipment. If the Surface Dose Rate measured value of a container is less than 200 mrem/h then that container should be managed as Contact Handled TRU (CH-TRU) waste.

NMED Characterization Methods

1. *Use AK to delineate waste streams, assign Hazardous Waste Numbers and Summary Category Group (SCG) to waste streams, and assign individual containers to waste streams.* Hazardous Waste Numbers and SCG are assigned on a waste stream basis; therefore the population of containers constituting a specific waste stream must be identified. AK must be used for the majority of this information. There are no other reasonable means (i.e., measurement) or confirmation techniques. For example, F codes depend on the generation process and therefore can only be identified by knowledge of the process used (part of AK). Because all allowable hazardous wastes (identified by Hazardous Waste Numbers) are managed identically at the WIPP, Hazardous Waste Numbers, or lack thereof, are not used for any operational management decisions. <u>Confirmation of Hazardous Waste Number assignment is therefore not required</u>.

2. The Characterization Method for establishing SCG is the same as described in the preceding NMED Characterization Method number 1.

3. *Use AK to demonstrate that the waste stream does not contain residual liquid, or that individual containers in a waste stream contain <1% residual liquid, or sample and NDEA/VE to determine the waste stream does not contain residual liquid, or NDEA/VE each container in a waste stream.* AK may provide definitive information (e.g., all liquids are eliminated by thermal process) to show that the waste stream contains no residual liquids. <u>If so, no confirmation is required</u>. Or AK may provide a record for each container

APPENDIX D

showing that there is <1% liquid. If so, no confirmation is required. If AK is not definitive, NDE or VE of a few representative containers may show that the waste stream does not contain residual liquids. If AK or sampling indicates that there may be some containers with >1% residual liquid, then NDE or VE of each container should be done.

4. *Use bounding analysis for RH contribution to reduce current HWFP Room Limits.* This is not a Characterization Method. This analysis has already been completed and can be found in Appendix 2 of the September 2001 draft Request for Class 3 Permit Modification (RH-TRU). This analysis bounded the possible contribution of VOCs from RH-TRU waste, and correspondingly lowered the Room Based Headspace VOC Limits set in the current HWFP. Reducing the Room Based VOC concentration limits eliminates the need to characterize RH-TRU waste for VOC concentrations.

Implementation: This column sets the standard of adequacy, sufficiency, or acceptability for the *Characterization Methods* selected in the adjacent column.

EPA Implementation

1. *WWIS electronically counts.* As pointed out in the corresponding Characteriza-tion Method column, this is a WIPP administrative action, not a characterization activity and therefore no adequacy standard is set.

2. *Use generator/storage site SCG designation from site documents.* If a site has previously determined a waste stream(s) to be one of three SCGs (S3000– homogeneous solids, S4000—soils and gravel, S5000—debris), then that determination is acceptable for program use. If the site has not determined the SCG and AK shows the process that generated the waste produces a uniform and homogeneous product, then designate the waste stream as S3000 or S4000 as appropriate, otherwise designate the waste stream as S5000. Current RH-TRU waste inventory information indicates that every retrievable RH-TRU waste stream that is a candidate for WIPP disposal already has a SCG designation. The associated WIPP administrative action is described in the Characterization Method.

3. *AK documents use of liquid management procedures, or a process that precludes liquids; otherwise assign 1%.* To assign 0%, AK is sufficient if it is documented that the waste stream generation procedures included liquid management procedures (e.g., procedures required the removal of liquids or the absorption of liquids), or it is documented that the waste stream generation process precludes liquids (e.g., material balance input records, thermal treatment, etc.). If this AK information is negative or unavailable, a default assignment of 1% is made.

4. *AK must include TRU information (sufficient to demonstrate waste stream concentration exceeds 100 nCi per gram), or the waste stream value must be established by sampling and measurement, or the TRU Activity must be measured for each container.*

- Because the same AK information is used for both the TRU Activity Characterization Objective and the TRU versus LLW determination, the more stringent TRU/LLW Characterization Objective sets the AK adequacy standard. Generally, AK is sufficient if the predominant TRU isotopes are identified (usually Pu238 or Pu239 and Am241), and there is waste stream activity information regarding those isotopes.
- If sufficient AK information is not available, TRU Activity for the waste stream can be established by measuring TRU Activity of representative containers. As with AK, only the predominant TRU isotope (usually Pu238 or Pu239 and Am241) need be measured. A sufficient number of containers shall make up the sample size in order to establish that the waste stream value exceeds 100 nCi per gram with 67% certainty.

APPENDIX D

- If a site elects to measure the TRU Activity of each container, measurement of the predominant TRU isotope is sufficient. If activity of the predominant isotope is marginally below 100 nCi per gram, a site may choose to include additional TRU isotopes in the measurement.

5. See Implementation item number 4 above.
6. *AK must include sufficient information with which to make an estimate of Total Activity for the waste stream, or the waste stream value must be established by sampling and measurement, or the Total Activity must be measured for each container.*

- AK is sufficient if the predominant high activity isotopes (e.g., Cs137) have been identified for the waste stream and there is a record of a few activity determinations for those isotopes.
- If sufficient AK information is not available, Total Activity for the waste stream can be established by measuring Total Activity of representative containers. A sample size of two to ten containers, depending on waste stream size, and measurement of only the predominant high activity isotopes, is sufficient to establish a Total Activity value for the waste stream.
- If a site elects to measure the Total Activity of each container, measurements of the predominant high activity isotopes are sufficient.

7. The analysis from Characterization Method eliminates any need for Implementation.
8. *Industry standard survey instruments.* This is a straightforward, common practice application of standard industry techniques to measure surface dose rate and requires no further explanation.
9. *Industry standard survey instruments.* This is a straightforward, common practice application of standard industry techniques to measure surface dose rate and requires no further explanation.
10. *Industry standard survey instruments.* This is a straightforward, common practice application of standard industry techniques to measure surface dose rate and requires no further explanation.

NMED Implementation

1. *Use generator/storage site waste stream designation, records of container assignments, Hazardous Waste Numbers if mixed waste, and SCG assignment* Every RH-TRU site planning to ship RH-TRU waste to the WIPP is currently operating under an EPA authorized State RCRA program. Waste streams have previously been designated under these programs, hazardous waste numbers have been assigned when appropriate, and the SCG for each waste stream has been determined.
2. Implementation for Physical Form is included in the above Implementation number 1.
3. *AK must document the use of liquid management procedures, or a process that precludes liquids, or there must be a record for each container indicating <1% residual liquid, or NDE/VE of a few representative containers from the waste stream can show the absence of residual liquids, or every container must undergo NDE/VE.*

- Liquid management procedures are standard operating procedures used in the waste generation process that eliminate or mitigate the presence of free liquids (e.g., removal or absorption). Examples of processes that preclude liquids are material input records that show no liquids were introduced to the process or a thermal treatment process that removes liquids.
- If sufficient AK information is not available, the absence of residual liquids in any

APPENDIX D

- container in a waste stream can be demonstrated with a statistical degree of certainty by NDE/VE of representative containers. A sufficient number of containers shall make up the sample size in order to establish that less than 10% of the containers in a waste stream contain more than 1% by volume residual liquids (with 67% certainty).
- A site may elect to NDE/VE every container in order to demonstrate that every container has less than 1% by volume residual liquid.

4. No Implementation is required for VOCs.

DQO: The RH-TRU Team applied the DQO process to each element of the proposed RH-TRU waste characterization program in order to ascertain whether formal DQOs need be developed. The DQO process was designed by the EPA primarily as a decision making tool applied to sampling and analysis studies for contaminated site remediation; however the process remains useful in the RH-TRU waste characterization program as a screening tool to determine if DQOs should apply. The seven-step DQO process is described in "Guidance for the Data Quality Objectives Process, EPA QA/G-4." Essentially DQOs establish what information is needed, and what level of quality the data should have associated with it, in order for a meaningful, risk based or consequence based decision to be made. "Data quality objectives (DQOs) for the data collection activity describe the overall level of uncertainty that a decision-maker is willing to accept in results derived from environmental data" (EPA SW-846). Using this process, the RH-TRU Team determined that there are no DQOs needed for the Characterization Objectives established for the RH-TRU waste characterization program.

EPA and NMED DQOs

- Step 1 of the DQO process is to "State the Problem." The problem statements for each element of the proposed RH-TRU waste characterization program are the Characterization Objectives listed in the first column of the attached Tables.
- Step 2 of the DQO process is to "Identify the Decision."

- There are no decisions to be made as a result of characterization activities for Table 1 EPA elements 1, 2, 3, 4, 6, and 9. These are all data collection activities to provide information for management, administrative, or load management decisions that may be needed in the future. Thus there are no DQOs for these elements.
- There are no decisions to be made as a result of characterization activities for Table 2 NMED elements 1, 2 and 4. Elements 1 and 2 are both administrative labeling activities done for the purpose of regulatory compliance that have no safety, repository performance, or technical consequences. NMED element 4 is potentially a data collection activity to enable future load management decisions regarding RCRA Permit *Room Based VOC Limits*. However the approach (Characterization Method from the Tables) taken for RH-TRU waste VOCs has eliminated this element as a RCRA characterization objective. Thus there are no DQOs for NMED elements 1, 2, and 4.

- Step 6 of the DQO process is to "Specify Tolerable Limits on Decision Errors."

- The decision(s) to be made for EPA elements 5, 7, and 8 are receive-not-receive at the WIPP, and for EPA element 10 the decision is receive as RH-TRU waste or CH-TRU waste. Legal and regulatory limits and/ or definitions drive decisions for each of these four elements, and there are no technical, safety, or repository performance consequences. Therefore "Tolerable Limits on Decision Errors" cannot be established for these four elements, and thus there are no DQOs for these elements.
- The decision to be made for NMED element 3 is receive-not-receive at the WIPP. Regulatory operational compliance drives this decision, and since there are no technical, safety, or repository performance consequences, no "Tolerable Limits on Decision Errors" can be established. Thus there is not a DQO for

> - <u>NMED element 3</u>.
>   <u>QAO</u>: The WIPP project has historically listed "accuracy," "precision," "representative-ness," "comparability," and "completeness" as the elements of QAOs. These are the same elements used in the EPA SW-846 (2.1) definition of DQOs. From an EPA perspective, <u>QAOs are synonymous with DQOs</u>, and thus the RH-Team established <u>no QAOs for either EPA or NMED Characterization Objectives</u> for the same reasons DQOs were not applicable.

NOTE: This copy has been edited for formatting purposes. The emphasis (italicized text) was added by DOE-Carlsbad Field Office. This table is part of the document DOE's Proposed Resolution and Action Items in Response to the Committee's Interim Report.
SOURCE: DOE (DOE-CBFO, 2002).

REFERENCE

DOE-CBFO. 2002. This material was submitted to the committee on January 18, 2002. This record is publicly available.

Appendix E

Information About Selected Transuranic Waste Generator Sites

The Department of Energy (DOE) provided this information in the March 2002 draft of both Document 1 (Notification of Proposed Change to the U.S. Environmental Protection Agency (EPA) Title 40 Code of Federal Regulations Part 194 Certification of the Waste Isolation Pilot Plant (WIPP), Attachment 3) and Document 2 (Request for RCRA Class 3 Permit modification to the NMED, Supplement 1). The committee did not verify or review this information. Table E.1 lists the Environmental Protection Agency (EPA) hazardous waste codes relevant to the sites mentioned in this appendix.

E.1 HANFORD SITE

The Hanford Site is one of the major contributors to the remote-handled transuranic (RH-TRU) waste inventory projected in WIPP (see Chapter 2).

E.1.1 Location and Description

The Hanford Site is located north of the Tri Cities (Richland, Kennewick, and Pasco), in Washington on a 1,450 km2 area of semiarid land in the Columbia River Basin in the southeastern corner of the State. Normal Columbia River elevations range from 119m, where the Columbia River enters the Hanford Site near the Priest Rapids Dam, to 104 m where it leaves the Hanford Site near the 300 area. Activities at Hanford are centralized in numerically designated areas. The reactor facilities are located along the Columbia River in what is known as the 100-Area. The reactor fuel processing and waste management facilities are in the 200-Area. The 300-Area, located adjacent to and north of Richland, contains the reactor fuel manufacturing facilities and the research and development laboratories. The 400-Area, 5 miles northwest of the 300-Area, contains the Fast Flux Test Facility, a sodium-cooled fast breeder reactor. The 600-Area covers all locations not specifically given an area designation. Adjacent to and north of Richland, the 1100-Area contains facilities associated with administration, maintenance, transportation, and materials procurement and distribution. The 3000-Area contains engineering and administrative offices. Administrative buildings, including the Federal Building, are located in the 700-Area, which is in downtown Richland. The Hanford Site is administered by the DOE Richland Operations Office.

APPENDIX E

E.1.2 Mission

The Hanford Site was acquired by the federal government in 1943 for the construction and operation of facilities to produce plutonium for the atomic weapons program during World War II. For more than 30 years, Hanford Site facilities were primarily dedicated to the production of plutonium for national defense and management of the wastes generated by chemical processing operations. In later years, programs at the Hanford Site became increasingly diverse, involving research and development for advanced reactors, renewable energy technologies, waste disposal technologies, and cleanup of contamination from past practices. The DOE has ended the production mission at the Hanford Site and is currently reorienting activities toward waste management and cleanup at the site. The mission now is one of environmental management, demonstration and application of advanced remediation technologies, and restoration of the Hanford Site.

E.1.3 Waste Information

Remote-handled TRU wastes from Hanford are the result of multiple sources. Some of these sources generated waste that now needs to be segregated between the categories of low-level waste, contact-handled transuranic (CH-TRU), and remote-handled transuranic waste. Some RH-TRU waste is currently stored in the 200-W Area in the Central Waste Complex and retrievably stored in the 200-E and 200-W areas in the Low-Level Burial Grounds. Future generation is forecast from activities related to the Hanford cleanup. The mission of the Central Waste Complex is to receive and store solid radioactive waste in a safe and environmentally compliant manner. The Central Waste Complex provides interim storage for mixed low-level waste, transuranic waste, and a small amount of low-level waste, awaiting treatment and final disposal. The design storage capacity is approximately 80,000 55-gallon drum equivalents; the operational capacity is 64,000 drum equivalents. The Central Waste Complex receives waste from both on-site and off-site waste generators.

Receipt of transuranic waste drums retrieved from the Low Level Burial Grounds began in 1999. All newly generated waste must meet acceptance criteria set by the Hanford Site Solid Waste Acceptance Program. Waste is generally packaged in 55-gallon drums unless alternate packages are dictated by size, shape, or other form of waste.

Each drum is handled individually using a hand truck, forklift, or crane. Drums are placed on wooden pallets with a maximum of four drums banded together; the pallets can then be stacked three high, or 12 drums per stack.

The storage buildings or pads have physical features that provide for segregated storage areas to maintain appropriate separation between groups of incompatible waste. Some RH-TRU waste will be segregated from the waste retrieved from the Low-Level Burial Grounds in the 200-Area. The waste to be retrieved from the Low-Level Burial Grounds is believed to primarily be low-level waste, though the waste will be segregated as it is retrieved and characterized. The RH-TRU wastes will be characterized and packaged at the M-91 facility. Another source of RH-TRU waste is the Plutonium Uranium Extraction Plant located in the 200-E Area. Products from this plant were weapons-grade plutonium, fuel-grade plutonium, depleted uranium, slightly enriched uranium, neptunium, and thorium. The plutonium conversion, process solution sampling, laboratory analyses, plant ventilation, and facility ventilation generated waste that was common to all these PUREX operations, including surgical gloves, plastic (polyvinyl chloride and polyethylene), tape, paper, glass, glovebox gloves, and metal tools and equipment.

The overall PUREX plant process consisted of seven fundamental interfacing processing units: feed preparation; solvent extraction, separation, and purification; solvent recovery and treatment for recycling; back-cycle waste system; acid recovery; waste treatment; off-gas treatment; and conversion of plutonium nitrate to plutonium oxide. Wastes generated were old failed equipment, port covers, replaced windows, brackets and hardware, and acid-soaked rags. Process solution sampling operations extract liquid samples from process control points to be used in system analysis and control. After use, all liquids were drained to the PUREX liquid waste system. Waste generated at this facility included broken glassware and plasticware, wet rags and paper, and piping and valving equipment. Laboratory Analysis Operations were performed on liquid process samples used for system analysis and control. After use, all liquids were drained to the PUREX liquid waste system. Again, wastes generated were broken glassware and plastic-ware, and wet rags and paper.

RH-TRU waste is also being stored in two spent nuclear fuel basins known as the K-Basins in Hanford's 100-K Area. This source consists of approximately 50 cubic meters of layered particulate material, which is generally called sludge. This sludge will be collected and transported to a stable interim storage location away from the Columbia River and eventually sent to a final disposal location. Sludge is found on the K-Basins, floors, in canisters, and in the basin pits. Several different types of sludge exist depending on the basin, canister type, and pit location where the particular sludge is found. Each type of sludge is a unique nonhomogeneous mixture possibly containing corroded fuel (i.e., uranium oxides, hydrates, hydride), cladding pieces, debris such as wind-blown sand or insects, rack and canister corrosion products, ion exchange resin beads, polychlorinated biphenyls, and/or fission products. In addition to the existing sludge material, other particulate materials are expected to be generated in the processing of fuel elements for dry storage. The sludge in the basins is commingled with spent nuclear fuel and is not considered a waste; however, when the sludge is separated from the spent nuclear fuel and removed from the basins, it will be dispositioned as RH-TRU waste.

For the purposes of differentiating spent nuclear fuel and debris from sludge, any material that is less than or equal to 0.64 cm (1/4 in.) in diameter is defined as sludge. Finally, RH-TRU waste will be generated from future activities related to facility stabilization and cleanup, maintenance of process equipment, laboratory operations, and Office of River Protection tank farm cleanup operations. This waste includes miscellaneous debris, equipment, and components and instrumentation trains removed during remediation of the high level waste tanks. The current estimates are that 207 cubic meters of RH-TRU is stored and an additional 944 cubic meters of waste will be generated. No canisterization of this waste has been performed with the majority of this waste to be generated (i.e., packaged) in the future.

The major radionuclides reported are cobalt-60, cesium/barium-137, strontium/yttrium-90, uranium-233, americium-241, plutonium-238, plutonium-239, plutonium-240, and plutonium -241. Based on known information about program activities, the estimated dose rate range is expected to be between 0.2 to 1,000 rem per hour. Due to the variety of processes, the presence of hazardous constituents will need to be ascertained on a stream-by-stream evaluation of the acceptable knowledge available on the generating process. Based on information previously reported in the TRU Waste Baseline Inventory Report, the following hazardous waste codes have been reported for RH-TRU wastes: F001, F002, F003, F004, F005, D001, D002, D005, D006, D007, D008, D009.

E.2 IDAHO NATIONAL ENGINEERING AND ENVIRONMENTAL LABORATORY

The Idaho National Environmental and Engineering Laboratory stores RH-TRU waste generated in national defense programs and research activities from across the country. For volume and radioactive waste inventory information, see Chapter 2.

E.2.1 Location and Description

The Idaho National Environmental and Engineering Laboratory is located in two primary areas: (1) the remote areas known as "the site" along the northern edge of the Snake River Plain in southeastern Idaho and (2) multiple locations southeast of the site in the city of Idaho Falls. Lying at the foot of the Lost River, Lemhi, and Bitterroot-Centennial Mountain ranges, the site covers nearly 2,300 square km (890 mi^2) of dry, cool desert. Most of the land withdrawn from public domain for use by DOE is undeveloped. The facilities located in Idaho Falls include administrative, scientific support, and nonnuclear research laboratories. During World War II, the U.S. Navy and U.S. Army Air Corps used a portion of the present site as a gunnery range. In 1949 the site was formally established as the National Reactor Testing Station, a place where the Atomic Energy Commission could build, test, and operate various types of nuclear reactors. Fifty-two reactors have been built at the Idaho National Engineering and Environmental Laboratory; of these, seven are operating or operable. The Radioactive Waste Management Complex encompasses 144 acres in the southwestern corner of the Idaho National Engineering and Environmental Laboratory. The Radioactive Waste Management Complex was established in 1952 as a controlled area for burial of solid radioactive wastes generated by Idaho National Engineering and Environmental Laboratory operations. In 1954 the burial ground was designated as a solid TRU waste disposal site. Until 1970, all TRU was buried below grade at the Radioactive Waste Management Complex. In November 1970 the Transuranic Storage Area was established for retrievable storage of waste contaminated with greater than 10 nanocuries of TRU alpha activity per gram of waste. In November 1976 the Intermediate Level TRU Storage Facility was established for retrievable storage of RH-TRU contaminated waste (greater than 200 millirem per hour). At the Intermediate Level TRU Storage Facility the radioactive waste is stored in above-grade vaults.

E.2.2 Mission

The Idaho National Engineering and Environmental Laboratory is a multi-program laboratory and has provided innovative technologies, defense-related support, and unique scientific and engineering capabilities to the nation. At present, areas of primary emphasis include nuclear reactor technology research and development, waste management and environmental restoration, advanced energy production and utilization technology development, defense-related support, technology transfer, and nonnuclear research and development projects. Development, transfer, and deployment of technologies to avoid and/or dispose of hazardous and/or radioactive waste and for remediation/restoration of previous disposal sites to protect the public, employees, and environment are also part of the Idaho National Engineering and Environmental Laboratory's mission.

E.2.3 Waste Information

RH-TRU waste generated in national defense programs and research activities from across the country was buried or stored at the Radioactive Waste Management Complex. The majority of the RH-TRU was generated at off-site facilities and shipped to the Idaho National Engineering and Environmental Laboratory for disposal or storage. These wastes came from Rocky Flats Environmental Technology Site, Argonne National Laboratory-East, Argonne National Laboratory-West, Battelle Columbus Laboratories, and other sites. The physical descriptions for these wastes are therefore consistent with the physical waste descriptions provided for each of these sites in different sections of this appendix, and the reader is directed there for those descriptions. In the case of Rocky Flats Environmental Technology Site, the RH-TRU was only differentiated from the CH-TRU shipped for disposal or storage at the Idaho National Engineering and Environmental Laboratory by its surface dose rate. The physical constituents of this TRU waste were typically the same and included cloth, paper, plastics, metals, rubber, sludge, or concrete. The TRU waste received at the Idaho National Engineering and Environmental Laboratory from November 1970 through July 1980 was placed on asphalt pads with an earthen cover to protect the waste from the environment until it could be permanently disposed of. Waste received after this time frame was placed in air-supported buildings for interim storage. The current estimate is that 84 cubic meters of RH-TRU is stored and that an additional 52 cubic meters of waste will be generated or recovered. No canisterization of this waste has been performed. Storage drums are typically 30 or 55 gallons. Major radionuclides reported include cobalt-60, cesium/barium-137, strontium/yttrium-90, uranium-235, americium-241, plutonium-238, plutonium-239, plutonium-240, and plutonium-241. Due to the wide variety of program activities, the estimated dose rate range is expected to vary from 0.2 to 100 rem per hour. Due to the variety of sources of RH-TRU at the Idaho National Engineering and Environmental Laboratory, the hazardous constituents could be varied, ranging from solvents and degreasers to metals, such as lead or cadmium. The presence of these hazardous wastes will need to be ascertained on a stream-by-stream evaluation of the acceptable knowledge available on the generating process at each generating sites. Based on information previously reported in the TRU Waste Baseline Inventory Report, the following hazardous waste codes have been reported for RH-TRU wastes: D008, D022, D028, D029, F001, F002, F003, and F005.

E.2.4 Supplemental Information

Additional potential sources of RH-TRU waste are being identified during review of Idaho National Engineering and Environmental Laboratory programs. One source is anticipated to be generated at the Idaho National Engineering and Environmental Laboratory from the Idaho Nuclear Technology and Engineering Center. This volume is anticipated to be about 900 cubic meters. The facility was previously known as the Idaho Chemical Processing Plant and was established in the early 1950s. Its original mission was to reprocess spent nuclear fuels by chemically separating out the reusable uranium and subsequently calcining the resultant high-level waste. The current mission is the storage of low-level, mixed low-level, and high-level waste and spent nuclear fuel and the development of treatment methods for high-level waste. Several evaluations are now in process to assess the treatment options for packaging the RH-TRU waste generated during processing of the high-level waste. This waste is expected to be classified as Waste Incidental to Reprocessing and disposed of at WIPP. Evaluations are also under way to examine treatment and packaging options for disposal. Depending on the option selected, the final packaged volume could be significantly greater than the current estimate of 900 cubic meters.

E.3 LOS ALAMOS NATIONAL LABORATORY

The Los Alamos Scientific Laboratory was established in 1943 by the U.S. Army's Manhattan Engineering District for the purpose of developing the first atomic weapons. Known as the Los Alamos Scientific Laboratory for many years, the name was changed to the Los Alamos National Laboratory in December 1980.

E.3.1 Location and Description

The Los Alamos National Laboratory is located approximately 97 km north of Albuquerque and 40 km west of Santa Fe in Los Alamos County, New Mexico. The laboratory facilities are dispersed among numerous technical areas spread over a 111-km^2 site on the Pajarito Plateau. The plateau consists of several finger-like mesas extending eastward from the Jemez Mountains to the Rio Grande Valley, with steep eroded canyons separating the mesas. The elevation of the mesas ranges from 1,890 to 2,377 m. Some Technical Areas are located in canyons.

E.3.2 Mission

Los Alamos National Laboratory was the research, development, engineering, design, and testing center for the Manhattan Project. The mission was the application of science and technology to problems of national security, including the maintenance of a strong defense, the fulfillment of arms controls commitments, and the guarantee of a secure energy supply for the future. Los Alamos National Laboratory's mission since World War II has included nuclear device design, research, development, testing, stockpile certification, and plutonium storage. Major programs currently include research in nuclear and conventional weapons development; nuclear fission and fusion; nuclear safeguards and security; verification and control technologies; fundamental research in particle physics, mathematics, chemistry, and materials; and waste management technology development and testing. Research on peaceful uses of nuclear energy has included space applications, power reactor programs, magnetic and inertial fusion, radiobiology, and medicine. Other programs include astrophysics, earth sciences, lasers, computer sciences, solar energy, geothermal energy, biomedical and environmental research, and nuclear waste management research.

E.3.3 Waste Information

RH-TRU waste is primarily generated by the Irradiated Materials Examination Group. This waste consists of solid wastes from laboratory and hot cell operations and includes small laboratory and hot cell equipment, materials such as fines resulting from grinding and cuttings, miscellaneous fuel specimens and their containers, and general debris from hot cell cleanup and decontamination. The general debris consists primarily of combustible waste, such as paper, rags, plastic, and rubber, with the plastic component consisting of tape, polyethylene and vinyl gloves, Tygon tubing, polystyrene, plastic vials, Teflon, and plexiglass. The cellulosic portion consists of rags, wood, cardboard, lab coats and coveralls, and paper. Noncombustible wastes that are present include cans, lids, graphite molds, furnaces, pyrochemical salts and related equipment, and glassware.

Some RH-TRU waste will be received from programmatic activities at the Sandia National Laboratory. The current estimate is that 98 cubic meters of RH-TRU is in storage and that 24 cubic meters of RH-TRU waste will be generated. The Los Alamos National Laboratory is unique in that it is the only site with waste packaged into canisters for handling and disposal at WIPP. Seventeen canisters of waste are currently packaged and stored in the waste storage area at Los Alamos National Laboratory. Other RH-TRU wastes are typically stored in small (1- to 5-gallon containers) and will be canisterized prior to disposal at WIPP. Major radionuclides reported include cesium/barium-137, strontium/yttrium-90, americium-241, plutonium-238, plutonium-239, plutonium-240, and plutonium-241. The estimated dose rate ranges are expected to be 1 to 100 rem per hour. Potential hazardous wastes included various cleaners and degreasers, cadmium, and lead. Based on information previously reported in the TRU Waste Baseline Inventory Report, the following hazardous waste codes have been reported for TRU wastes at Oak Ridge National Laboratory: F001, D006, D008.

E.4 OAK RIDGE NATIONAL LABORATORY

The Oak Ridge National Laboratory is the site that stores most of the existing RH-TRU waste in DOE's inventory (see Chapter 2).

E.4.1 Location and Description

The Oak Ridge National Laboratory is located 10 miles southwest of downtown Oak Ridge, Tennessee, and 32 km northwest of Knoxville, Tennessee. The Oak Ridge National Laboratory site occupies about 10,000 acres of the 35,252-acre-Oak Ridge Reservation. The site covers portions of both Melton and Bethel valleys. Approximately 1,100 acres in the Melton and Bethel valleys has been developed. The Oak Ridge National Laboratory is under the auspices of the DOE/Oak Ridge Operations Office, which supports production of nuclear weapon components for national defense programs, production of enriched uranium for defense requirements and for fueling nuclear power plants, processing of uranium feed materials for DOE's plutonium production reactors, and extensive energy research and development in all DOE program areas.

E.4.2 Mission

The Oak Ridge National Laboratory was established in 1942 in support of the Manhattan Project. The primary mission of the Oak Ridge National Laboratory has been to carry out applied research and engineering development in fission, fusion, and other energy technologies and to conduct scientific research in basic physical and life sciences. Relevant missions include isotope production and processing, research and development, waste management, and decontamination and decommissioning of operating units, and advanced reactor development work. In addition, the Oak Ridge National Laboratory conducts several activities for DOE defense programs. The principal nonweapons-related activities include nuclear power development and magnetic fusion research.

E.4.3 Waste Information

Most of the RH-TRU waste at Oak Ridge National Laboratory has been generated as a result of special isotope separation activities. The Radiochemical Engineering Development Center at Oak Ridge National Laboratory has processed americiumcurium targets that were irradiated in the High Flux Isotope Reactor to produce higher actinides such as californium, berkelium, einsteinium, and fermium. The ion exchange processing of these targets generated and consists of two distinct waste streams, a solid stream (debris waste) and a sludge/liquid stream. Solid RH-TRU debris waste consists primarily of miscellaneous hot cell waste (e.g., paper, glass, plastic tubing, shoe covers, wipes), high-efficiency particle activated filters from off-gas cleanup systems, and discarded equipment (e.g., chemical processing racks, vacuum pumps). This waste was contaminated with the original target material and the higher actinides, except for the einsteinium and fermium, which decayed quickly due to their very short half-lives. This solid debris is stored in concrete casks as RH-TRU waste. There is substantial historical process knowledge about the content of each cask (ORNL, 1989). For example, some casks contain a large number of 1 gallon pails of hot cell waste while others may have a single failed piece of equipment. The unshielded individual waste packages in the casks typically have radiation levels that measure 10 to 2,000 rem per hour; the majority are below 100 rem per hour.

Hazardous wastes in RH-TRU solid waste primarily consist of lead that was used as shielding and limited amounts of mercury from discarded mercury vapor lamps. The RH-TRU solid waste is typically contained in cylindrical concrete casks 1.4 m (4.5 ft.) in diameter by 2.3 m (7.5 ft.) high. Wall thickness of the casks are currently either 15.2 or 30.5 cm (6 or 12 in.) thick, depending on the radiation level of the contents.

The majority of RH-TRU sludges at the Oak Ridge National Laboratory are the result of waste accumulation from the past 50 years of liquid waste operations there. These sludges are residuals from sluicing operations conducted in the early 1980s when the majority of the inactive gunite tank contents were removed for hydrofracture disposal at Oak Ridge National Laboratory. This liquid waste is now stored in the Melton Valley Storage Tanks. RH-TRU sludges continue to accumulate due to on going research and development programs, that produce transuranic isotopes for medical, industrial, and government applications.

The surface dose rates of these sludges are generally near 10 rem per hour (unshielded). The current estimate is that 1,308 cubic meters of RH-TRU is stored and that an additional 534 cubic meters of waste will be generated. No canisterization of this waste has yet been performed, but canisterization is scheduled to begin in fiscal year 2003 and will be performed by Foster-Wheeler, the contractor performing the characterization and packaging activities. Major radionuclides reported include cobalt-60, cesium/barium-137, strontium/yttrium-90, europium-152, europium-154, uranium-233, uranium-235, uranium-238, americium-241, plutonium-238, plutonium-239, plutonium-240, plutonium-241, curium-244, and californium-252. The estimated dose rate range is expected to be 0.2 to 1,000 rem per hour. Based on information previously reported in the TRU Waste Baseline Inventory Report, the following hazardous waste codes have been reported for RH-TRU wastes at Oak Ridge National Laboratory: D006, D007, D008, D009, D011.

E.5 ENERGY TECHNOLOGY ENGINEERING CENTER

The Energy Technology Engineering Center is a small generator site. This site will serve as model to other small generator sites in the implementation of the RH-TRU waste characterization plan. The site-specific RH-TRU waste characterization implementation plan from this site will be submitted to the regulatory agency along with Documents 1 and 2 (see Chapters 4 and 5).

E.5.1 Location and Description

The Energy Technology Engineering Center occupies 90 of 290 acres of land shared with the Santa Susana Field Laboratory. The Santa Susana site consists of a total of 2,700 acres located in the Simi Hills of Ventura County, California, approximately 48 km northwest of downtown Los Angeles. The facilities include former fuel fabrication facilities, a hot cell, a reactor test building, a storage vault, an on-site transport cask, and other radiologically contaminated support laboratories and areas.

E.5.2 Mission

Energy Technology provides facilities for the testing of equipment, materials, and components for nuclear and other energy programs. Components include steam generators, pumps, valves, instrumentation, and other support elements for power plant design. Various types of testing include reliability, seismic, and performance demonstrations. Current activities include non-nuclear testing and cleanup and environmental restoration from prior nuclear testing programs, such as decontamination and decommissioning of a hot cell licensed by the Nuclear Regulatory Commission that was used for DOE activities.

E.5.3 Waste Information

The RH-TRU waste at Energy Technology Engineering Center was generated during DOE fuel decladding and decontamination and decommissioning operations. The RH-TRU waste consists of two waste streams: (1) hot laboratory drain line residue and (2) a single drum of debris waste from multiple sources. The drain line residue is currently stored in 28 concrete-shielded 55-gallon drums and one 30-gallon drum. An additional amount of about 22 gallons of sludge is estimated to be in a 3,000-gallon drain tank and about 10 gallons of residue in two weir boxes. The total volume of the unpackaged waste is about 0.3 cubic meters. The total volume of the drain line residue material when all of the material is repackaged for on-site storage in concrete-shielded drums is estimated to be forty 55-gallon drums plus one 30-gallon drum, or about 8.5 cubic meters. The single 55-gallon drum from the debris waste has a total volume of 0.21 cubic meters. The total RH-TRU volume is therefore 8.7 cubic meters. There are no plans to canisterize RH-TRU waste at the Energy Technology Engineering Center. The RH-TRU waste will be sent to an intermediate site for final waste disposal characterization and final packaging. It is anticipated that the intermediate site may either be Oak Ridge or Hanford. Major radionuclides expected include cobalt-60, cesium/barium-137, strontium/yttrium-90, americium-241, plutonium-239, plutonium-240, and plutonium-241. Dose rates are relatively low in the range of 0.2 to 10 rem per hour. Based on acceptable knowledge documentation, such as process records, and supplemental analyses, the Energy Technology Engineering Center has assigned waste codes below to these waste streams. Recent tests have also shown that the drain line residue stream has polychlorinated biphenil levels approaching 100 ppm. The Energy Technology Engineering Center plans to manage its RH-TRU waste as three separate waste streams. Based on information previously reported in the TRU Waste Baseline Inventory Report, the following hazardous waste codes have been reported for RH-TRU wastes at this site: D008 and D009.

E.6 BATTELLE COLUMBUS LABORATORIES

This site is the only one actively characterizing its RH-TRU waste inventory. This site is seeking authorization to ship its RH-TRU waste to the Hanford Site for interim storage (see Chapter 2).

E.6.1 Location and Description

The Battelle Columbus Laboratories consists of two research complexes: one at 505 King Avenue in the city of Columbus, Ohio, and the second, the West Jefferson Site in Madison County west of Columbus. The King Avenue facility houses corporate offices and general research laboratories. The West Jefferson site consists of a number of facilities formerly dedicated to nuclear research. The King Avenue facility is located in the western central portion of the city of Columbus. The 10-acre complex accommodates 21 buildings and is bounded on the north by King Avenue, on the east by Battelle Boulevard, on the south by Fifth Avenue, and on the west by the Olentangy River. The Columbus campus of Ohio State University lies immediately north across King Avenue. The remaining contiguous area is a moderately dense residential neighborhood. The West Jefferson Site is located in West Jefferson, Ohio, approximately 24 km west of the King Avenue facility. The 1,000-acre tract accommodates 21 buildings in the Engineering Area, Experimental Ecology Area, and Nuclear Services Area. The site boundary on the north is about 1 mile south of I-70, on the east is Big Darby Creek, on the south are the Conrail tracks, and on the west is the Georgeville-Plain City Road. The land to the north, west, and south for 2 miles is cleared farmland and/or wood lots.

E.6.2 Mission

The mission of Battelle in 1943 was to perform atomic energy research and development activities for the Manhattan Engineering District. Since that time Battelle has continuously performed research and development at these facilities. Past programs have included uranium ore processing and benefaction studies, metallurgical and ceramic process development, corrosion studies, fabrication of weapons components, ballistics experiments, hot cell work, critical assembly and criticality experiments, and an experimental reactor.

E.6.3 Waste Information

The main DOE-sponsored work currently being done at Battelle Columbus Laboratories is decontamination and decommissioning of the contaminated buildings and hot cells at the West Jefferson location. This work is being performed under the direction of the Battelle Columbus Laboratories Decommissioning Project. The decontamination and decommissioning activities involve removing from the hot cells equipment used in the fuel examination process, materials such as fines resulting from grinding and cuttings, miscellaneous fuel specimens and their containers, and general debris. The current estimate is that a total of 20.8 cubic meters of RH-TRU waste will be generated; some of this waste is currently in the process of being packaged. The proposal is to consolidate this waste at an intermediate facility for disposal characterization and any additional packaging. A memorandum of agreement is currently under development for use of the Hanford Reservation as the intermediate site. None of the RH-TRU waste is canisterized for transport in the RH-72B cask. The current plan is to use the CNS 10–160B to ship the majority of waste; therefore, the RH-TRU waste is being packaged in 55-gallon drums. Major radionuclides reported include cobalt-60, cesium/barium-137, strontium/yttrium-90, americium-241, plutonium-239, plutonium-240, and plutonium-241. Measured dose rates range from 0.2 to 150 rem per hour. Based on acceptable knowledge documentation (e.g., production and process records, analytical lab records), the following hazardous waste codes have been reported for RH-TRU wastes: D005, D007, D009, D011, F001, F002, and F005.

TABLE E.1 EPA Hazardous Waste Codes Relevant to this Appendix

Hazardous Waste Code	Criteria and Characteristics of Hazardous Waste
D001	Ignitability
D002	Corrosivity
D003	Reactivity
D004–D029	Toxicity: A solid waste whose extract under the test procedure specified under 40 Code of Federal Regulations Part 261.24 contains one or more constituents at concentrations greater than those specified in the Maximum Concentration of Contaminants for the Toxicity Characteristic Table. The toxicity characteristics of selected hazardous waste codes together with their maximum concentration of contaminants are listed below (D004–029).
D004	Arsenic >5 mg/L
D005	Barium >100 mg/L
D006	Cadmium >1 mg/L
D007	Chromium >5 mg/L
D008	Lead >5 mg/L
D009	Mercury >0.2 mg/L
D011	Silver >5 mg/L
D022	Chloroform >6 mg/L
D028	Dichloroethane (1,2-) >0.5 mg/L
D029	Dichloroethylene (1,1-) >0.7 mg/L
	Hazardous Waste from Non Specific Sources
	The following spent halogenated solvents used in degreasing:
F001	Tetrachloroethylene, trichloroethylene, methylene chloride, 1,1,1- trichloroethane, carbon tetrachloride, and chlorinated fluorocarbons; all spent solvent mixtures/blends used in degreasing containing, before use, a total of 10% or more (by volume) of one or more of the above-halogenated solvents or those solvents listed in F002, F004, and F005 and still bottoms from the recovery of these spent solvents and spent solvent mixtures.
	The following spent halogenated solvents:
F002	Tetrachloroethylene, methylene chloride, trichloroethylene, 1,1,1-trichloroethane, chlorobenzene, 1,1,2-trichloro-1,2,2-trifluoroethane, ortho-dichlorobenzene, trichlorofluoromethane, and 1,1,2-trichloroethane; all spent solvent mixtures/blends containing, before use, a total of 10% or more (by volume) of one or more of the

APPENDIX E

	above-halogenated solvents or those listed in F001, F004, or F005 and still bottoms from the recovery of these spent solvents and spent solvent mixtures. The following spent non halogenated solvents:
F003	xylene, acetone, ethyl acetate, ethyl benzene, ethyl ether, methyl isobutyl ketone, N-butyl alcohol, cyclohexanone, and methanol; all spent solvent mixtures/blends containing, before use, only the above spent non halogenated solvents; and all spent solvent mixtures/blends containing, before use, one or more of the above non halogenated solvents, and a total of 10% or more (by volume) of one or more of those solvents listed in F001, F002, F004, and F005; and still bottoms from the recovery of these spent solvents and spent solvent mixtures. The following spent non halogenated solvents:
F004	Cresols and cresylic acid, and nitrobenzene; all spent solvent mixtures/blends F004 containing, before use, a total of 10% or more (by volume) of one or more of the above non halogenated solvents or those solvents listed in F001, F002, and F005; and still bottoms from the recovery of these spent solvents and spent solvent mixtures. The following spent non halogenated solvents:
F005	Toluene, methyl ethyl ketone, carbon disulfide, isobutanol, pyridine, benzene, 2-ethoxyethanol, and 2-nitropropane; all spent solvent mixtures/blends containing, before use, a total of 10% or more (by volume) of one or more of the above non halogenated solvents or those solvents listed in F001, F002, or F004; and still bottoms from the recovery of these spent solvents and spent solvent mixtures.

REFERENCE

ORNL (Oak Ridge National Laboratories). 1989. L.S.Dickerson, Remote-Handled Transuranic Solid Waste Characterization Study: Oak Ridge National Laboratory, ORNL/TM-11050. Oak Ridge National Laboratory. Oak Ridge, Tenn.

Appendix F

Overview of the Contact-Handled Transuranic Waste Characterization Plan

The characterization program described here has been developed for contacthandled transuranic (TRU) waste and applied to date to TRU mixed waste. The methods, equipment, procedures, determination of uncertainty, and other protocols used at DOE sites to perform these characterizations have been approved by the Department of Energy (DOE) Carlsbad Field Office, New Mexico Environment Department (NMED) and EPA. The major procedures are described in the following sections:

F.1 DETERMINATION OF THE ORIGIN AND COMPOSITION OF THE WASTE BY ACCEPTABLE KNOWLEDGE

Acceptable knowledge (AK) of the origin and composition of the waste must be documented to provide evidence that the waste has a defense origin (by the terms of the Land Withdrawal Act, only defense-related TRU waste may legally be sent to the Waste Isolation Pilot Plant (WIPP) and to provide characterization information on the waste constituents. The DOE Carlsbad Area Office and EPA use the AK documentation to certify each "waste stream" (i.e., waste-generating process), and TRU waste sent to WIPP must come from a certified waste stream.

F.2 SAMPLING AND ANALYSIS OF HOMOGENEOUS WASTE FOR RESOURCE CONSERVATION RECOVERY ACT CONSTITUENTS

Most of the TRU waste is heterogeneous in nature and requires no further characterization beyond AK to satisfy the regulatory requirements of Resource Conservation Recovery Act (RCRA). For homogeneous waste a fraction of the waste containers (e.g., 55-gallon drums or standard waste boxes) are cored to extract representative samples that are analyzed for constituents (e.g., volatile and semivolatile organic compounds, toxic metals other hazardous chemicals) regulated by RCRA. Both the AK procedure (for heterogeneous waste) and the sampling and analysis procedure (for homogeneous waste) were proposed by DOE for the terms of operation that would be specified in its RCRA permit. These terms have been accepted by New Mexico, which was delegated authority by EPA to regulate RCRA materials and mixed waste and to issue the RCRA Part B permit in October 1999.

F.3 RADIOGRAPHY

A radiography procedure using X rays, also called real-time radiography, is performed on all waste containers to look for items such as pressurized cans or free-standing liquids that are prohibited from being transported under U.S. Department of Transportation regulations. If any of these items are present in a waste container, the prohibited materials are removed and the contents repackaged. This radiographic examination is also used to confirm the AK characterization information.

F.4 VISUAL EXAMINATION

A visual examination is performed on a fraction of the waste containers by placing the waste contents into a glovebox to verify the AK and real-time radiography information. DOE proposed that 2 percent of the initial population of containers of each waste stream be visually examined, and if these evaluations resulted in few miscertifications, the percentage of subsequent waste containers to undergo visual examination would be reduced. In October 1999, New Mexico in its RCRA permit stipulated the initial fraction of containers to undergo visual examination to be 11 percent.

F.5 RADIOASSAY AND DETERMINATION OF FISSILE ISOTOPE CONTENT

The number of curies of each transuranic isotope is determined by radioassay (e.g., gamma scans) to a specified precision and accuracy. The fissile isotope content is assessed using non-destructive assay methods, such as passive-active neutron systems. This information is used to meet the Nuclear Regulatory Commission requirement restricting the amount (several hundred grams) per container of each fissile species to ensure criticality safety.

F.6 HEADSPACE GAS SAMPLING

Headspace gas sampling is carried out on all waste containers for flammable gases (specifically, volatile organic compounds, hydrogen, and methane). This procedure has been proposed as a means of checking on conformity with the U.S. Department of Transportation regulations, such as Title 40 Code of Federal Regulations (CFR) Part 173 and Title 40 CFR Part 177, and U.S. Nuclear Regulatory Commission regulations, such as Title 10 CFR Part 71, that address the transport of flammable and/or gas-generating substances with radioactive materials. DOE has proposed the headspace gas sampling procedure in its application to the U.S. Nuclear Regulatory Commission for a licensing certificate on the transportation package (named the TRansUranic PACkage Transporter, or TRUPACT-II) that is loaded with waste containers for transport by truck to WIPP.

Appendix G

Non-Destructive Techniques for Remote-Handled Transuranic Waste Characterization

This appendix is based on information gathered by the committee in the time frame allowed by this study. It should not be considered a comprehensive review of all techniques being considered to characterize remote-handled transuranic (RH-TRU) waste.

Non-destructive characterization of RH-TRU waste is challenging because of the high surface dose rate of waste canisters (up to 1,000 rem per hour) resulting from the high background gamma and neutron radiation generated by fission and activation products in the waste. To reduce the risk of worker exposure during characterization of RH-TRU waste, non-destructive examination and assay techniques are needed to avoid performing destructive analysis such as visual examination and radiochemical assays. According to the information gathered from DOE's Transuranic and Mixed Waste Focus Area: "Non-destructive examination and assay techniques to characterize remote-handled wastes must still be developed and demonstrated" (DOE-ID, 2001).

Non-destructive assay (NDA) identifies radioactive components, and non-destructive examination (NDE) determines the physical makeup of the waste. These non-destructive technologies must be capable of accurately identifying the physical and radiological properties of wastes in lead-lined containers as well as correcting for precision or bias problems caused by matrix effects (such as heterogeneity problems and interferences with heavy metals in waste) and by the high radiation background. In particular, the neutron emitting RH-TRU solid debris waste at Oak Ridge National Laboratory is the most challenging to characterize.

The following is a brief overview of the major non-destructive characterization techniques currently under development by various Department of Energy (DOE) generator sites, contractors, and private companies. Further development of these techniques and demonstration on actual RH-TRU waste is pending the outcome of the characterization plan for RH-TRU waste. The major NDA/NDE techniques considered for RH-TRU waste are the following:

G.1 RADIOGRAPHY

This NDE technique is often referred to as X-ray radiography or real-time radiography (RTR). X-rays provide a direct measure of material density and as such can be used to establish the physical form of the waste and identify certain prohibited items. In addition, operators can be trained to look for indicators of prohibited items such as electrical components that could contain polychlorinated biphenyls. This technique consists of an X-ray source combined with a TV camera attached to an X-ray detector.

Roney and White (2001) studied the characterization of RH-TRU and lead-lined drums using X-ray imaging techniques. (Roney and White, 2001). Although no real RH-TRU waste was used in their work, they demonstrated the applicability of radiography to surface dose rates up to 100 rem per hour. TMFA states that: "current real time radiography (RTR) techniques cannot analyze the content of the lead-lined drums. Similarly, RH waste inserts and casks are too thick for conventional RTR analysis and interrogation may be further complicated by high radiation levels and contributed 'white noise' from inside the drum" (DOE-TMFA, 2002a).

To the best of the committee's knowledge, radiography has not been demonstrated on real or surrogate waste with surface dose rates close to 1,000 rem per hour.

G.2 GAMMA-RAY SPECTROSCOPY

Gamma-ray spectroscopy uses the gamma emission and transmission properties of radionuclides contained in the waste to obtain qualitative and quantitative information of radionuclide content. Gamma-ray tomography is a variation of gamma spectroscopy and is used to determine the spatial and quantitative information of radionuclide distribution in waste containers. Matrix interferences are corrected by scanning for known gamma-ray sources. For further information on related technologies, see, for example, Meeks and Chapman (1997).

G.3 GAMMA-RAY SPECTROSCOPY COMBINED WITH ACCEPTABLE KNOWLEDGE

The GSAK approach is used to determine the quantities and types of radionuclides in waste drums. This technique is based on measurement of the radionuclide distribution in the waste followed by normalizing this distribution to the activity of a gamma-emitter such as cesium-137 or cobalt-60, which can be measured outside the waste drum. Based on the measured surface dose rates for cesium-137 or cobalt-60, the amount of these two radionuclides inside the drum can be estimated. This in turn allows the activity of the other radionuclides in the waste drum to be estimated using their ratio to cesium-137 or cobalt-60. The radionuclide distribution may be determined through gamma spectroscopy methods or other nuclear counting methods for alpha- and beta-emitters. Because alpha- and beta-emitters are often difficult to measure in waste, measurements are sometimes used in conjunction with detailed computer modeling to develop a radionuclide distribution that includes these hard-to-measure radionuclides. This technique requires an accurate knowledge of the waste to be characterized before measurements are integrated with computer modeling.

The Battelle Columbus Laboratories have demonstrated this technique with actual RH-TRU waste (Biedscheid et al., 2002). The Idaho Engineering and Environmental Laboratory is also planning to use this technique for its RH-TRU waste originated in Argonne National Laboratory-East (Bhatt and Clements, 2001). For further information about this technique and its demonstrations see Hartwell et al. (1997, 2000), Jensen (2001), and Klann and Grimm (2000).

G.4 PASSIVE AND ACTIVE NEUTRON MEASUREMENT

Passive and active neutron measurements are widely used in waste management and safeguard operations for fissile mass measurements (usually plutonium-239, plutonium-241, or uranium-235) and for determination of alpha contents for disposal sites. This technique is based on the spontaneous fission rate determination of fissile material present in the container in (passive neutron measurement mode) and on the induced fission rate by neutron interrogation (active neutron measurement mode). The technique is sensitive to extraneous neutrons coming from secondary alpha decays and from cosmic background. This technique was used to characterize 10 actual RH-TRU waste containers and over 200 containers of surrogate RH-TRU waste at Los Alamos National Laboratory (Estep et al., 1989; Estep, 2001, 2002). For further information on this technique and its variations see DOE (1998), Ensslin et al. (2000), Royce and Lucero (2001), and Schultz et. al. (1995).

G.5 MULTI-DETECTOR ANALYSIS SYSTEM

The multi-detector analysis system (MDAS) technique is used to obtain a direct isotopic measurement with no prior knowledge of the waste. MDAS interrogates the (unshielded) waste with neutrons, and then collects energy coincidence measurements through multiple detectors for both gamma and neutron data over a small period of time. Coincident neutrons provide information on the quantity of fissile material present, and coincident gamma rays provide information on specific isotopes from fission products. This technique is designed to reduce background measurement interferences so that a direct isotopic measurement is provided. Uncertainties associated with current passive/active neutron measurement techniques are reduced by measuring the coincidence energies fast neutrons. A demonstration of this technique on RH-TRU waste was scheduled for 2001 but did not take place due to problems with the external neutron source. For further information on this technique, see DOE-TMFA (2002b), DOE-ID (2001) and Shelton-Davis (2001).

REFERENCES

Bhatt, R. and T.Clements, Jr. 2001. Proposed radiological characterization approach for fuel-based remote-handled waste stored at the Idaho National Engineering and Environmental Laboratory. Paper presented at the Non-Destructive Assay Interface Working Group Meeting. July 20. Indian Wells, Calif.

Biedscheid, J., S.Stahl, M.Devarakonda, K.Peters, and J.Eide. 2002. Adequacy of a Small Quantity Site RH-TRU Waste Program in Meeting Proposed WIPP Characterization Objectives. In Proceedings of the WM'02 Conference. February 24– 28. Tucson, Ariz.

DOE. 1998. Non-destructive Waste Assay Using Combined Thermal Epithermal Neutron Interrogation. Innovative Technology Summary Report. Mixed Waste Focus Area DOE/EM-0465. U.S. Department of Energy. Washington, D.C. Available at: <http://apps.em.doe.gov/ost/pubs/itsrs/itsr1568.pdf>.

DOE-ID. 2001. Transuranic and Mixed Waste Focus Area Multi-Year Program Plan FY2001. DOE/ID-10659. U.S. Department of Energy, Idaho Operations Office. Washington, D.C. Available at: <http://tmfa.inel.gov/Documents/MYPP01.pdf>.

DOE-TMFA. 2002a. STCG Need and Technical Response. Need Title: Techniques to Analyze Shielded and Remote-Handled Drums. MW01-CB-01–02. Available at: <http://tmfa.inel.gov/Needs/needcomp.asp?wp=MW01&id=178>.

DOE-TMFA. 2002b. TMFA TTP Summary Information. Available at: <http://tmfa.inel.gov/Needs/ttpscope.asp?id=156>.

APPENDIX G

Ensslin, N., D.R.Mayo, M.C.Browne, L.A.Carrillo, L.A.Foster, W.H.Geist, J.E. Stewart, and K.Veal (Los Alamos National Laboratory). 2000. Evaluation of Several Fast Neutron Coincidence Counter Options for Characterization of Remote-Handled Transuranic Waste. Seventh Non-destructive Assay Waste Characterization Conference. May 22–26. Salt Lake City, Utah.

Estep, R.J., K.L.Coop, T.M.Deane, and J.E.Lujan. 1989. A Passive-Active Neutron Device for Assaying Remote-Handled Transuranic Waste. LA-UR-89–3736. Paper presented at the Topical Meeting on Non-destructive Assay of Radioactive Waste. November 17–22. Cadarache, France.

Estep, R. 2001. LANL Experience Assaying RH-TRU Waste Using a Modified DDT/PAN Counter. Presented at the Non-destructive Assay Interface Working Group Meeting. July 20. Indian Wells, Calif.

Estep, R. 2002. Personal communication with National Research Council staff, May 15.

Hartwell, J.K., W.Y.Yoon, and H.K.Peterson (Idaho National Engineering and Environmental Laboratory) . 1997. A Preliminary Evaluation of Certain NDA Techniques for RH-TRU Characterization. Proceedings of the Fifth NDA/NDE Waste Characterization Conference. Salt Lake City, Utah. Available at: <http://plutoniumerl.actx.edu/thefifth.html>.

Hartwell, J.K., R.T.Klann, and M.E.McIlwain. 2000. Gamma-Ray Spectrometric Characterization of Overpacked CC104/107 RH-TRU Wastes: Surrogate Tests. Proceedings of the Seventh Non-destructive Assay Waste Characterization Conference. May 22–26. Salt Lake City, Utah.

Jensen, C. (Battelle Columbus Laboratories). 2001. Methodology for Determining Radiological Properties of RH-TRU Waste. Paper Presented at the Non-Destructive Assay Interface Working Group Meeting. July 20. Indian Wells, Calif.

Klann, R.T. and K.N.Grimm. 2000. Acceptable Knowledge for Argonne National Laboratory Remote-Handled Transuranic Waste. Proceedings of the Seventh Non-destructive Assay Waste Characterization Conference. May 22–26. Salt Lake City, Utah.

Meeks, A.M., and J.A.Chapman. 1997. Development of the Remote-Handled Transuranic Waste Radioassay Data Quality Objectives: An Evaluation of RH-TRU Waste Inventories, Characteristics, Radioassay Methods and Capabilities. ORNL/TM-13362. Oak Ridge National Laboratory. Oak Ridge, Tenn.

Roney, T.J., and T.A.White. 2001. Characterization of RH-TRU and Lead-Lined Drums Using X-Ray Imaging Techniques. INEEL/EXT-2001–00625. Idaho National Engineering and Environmental Laboratory. Idaho Falls.

Royce, R., and R.Lucero (BNFL Instruments, Inc.). 2001. Characterization of RH Waste at Melton Valley Using PAN/GEA/ETA Technology. Available at: <http://tmfa.inel.gov/Documents/RHWASTE.pdf>.

Schultz, F.J., G.Vourvopoulos, P.C.Womble, and M.L.Roberts. 1995. A feasibility study for a LINAC-based transuranic waste characterization system. Journal of Radioanalytical and Nuclear Chemistry, 193(2):369–375. Available at: http://www.wku.edu/Dept/Academic/Ogden/Phyast/API/research/hawaii94.pdf>.

Shelton-Davis, C.V. 2001. Multi-Detector Analysis System Fiscal Year 2000 Summary Report. March. INEEL/EXT-01–00367. Idaho National Engineering and Environmental Laboratory. Idaho Falls.

Appendix H

Waste Dose Rates and Characterization Cost Estimates

The committee provides the following worker dose and characterization cost estimates as examples of the type of assessments that could be presented in the three implementation plans (from Battelle Columbus Laboratories, Energy Technology Engineering Center, and Oak Ridge National Laboratory) that the U.S. Department of Energy (DOE) is planning to submit to the U.S. Environmental Protection Agency and the New Mexico Environment Department along with its characterization plan for remote-handled transuranic (RH-TRU) waste. The committee's assessment is based on the information gathered during the study. It is not to be considered complete and is not meant to replace DOE's own assessment of worker doses and costs.

The predominant exposure pathways for workers from routine waste handling activities are direct exposure to external penetrating radiation or inhalation from surface contamination, and inhalation of airborne radioactive material generated during characterization and handling activities. The radiogenic health effects associated with TRU waste are due primarily to alpha and gamma radiation. Beta particles and neurons may also pose health concerns for RH-TRU waste if the container is breached or waste is exposed during handling or characterization.

H.1 COMMITTEE'S EXAMPLE OF ESTIMATE FOR WASTE DOSE RATES

Because of the high surface dose rates, all of the handling operations involving RH-TRU waste are performed in hot cells, where available, and/or by remote-handling equipment. Good radiological safety practices require that all radiological activities must maintain worker dose levels ALARA (as low as reasonably achievable). In the absence of appropriate radiological safety measures, the potential for significant radiation exposure is high. The estimated volume-weighted mean RH drum surface dose rate at individual generator sites is 470 mSv per hour (47,000 mrem per hour) (Wu, 2001). Therefore, the cost associated with the handling of RH-TRU waste may be substantial.

Doses from handling RH-TRU waste are expected to be less than contact-handled transuranic (CH-TRU) waste due to requirements for remote handling. For instance, the DOE-Carlsbad Field Office estimated for the period 1999–2001 that the collective dose for handling CH-TRU waste at the Waste Isolation Pilot Plant (WIPP) is approximately 0.08 person-mSv (0.008 person-rem) per transportation cask type TRUPAC-II. According to Wu (2001), this is about two times higher than the estimated collective dose to workers handling a RH72-B type cask of RH-TRU waste.

Doses to workers depends on several factors including the number of drums characterized and handled, types of sampling and measurements conducted, number of workers involved in the characterization and drum handling, worker experience and training, and facilities, equipment and shielding. Table H.1 compares estimated collective doses to workers at three different sites. Because these data are from different sites they may not be comparable.

TABLE H.1 Committee's Estimates of Worker Doses at Selected Sites

Site	Collective Dose	Comment
Battelle Columbus Laboratories	1.0 person-mSv (100 person-mrem) per container	This dose corresponds to an average of five workers handling the containers and includes waste characterization and handling containers for storage.[1]
Argonne National Laboratory-East	0.22–0.64 person-mSv (22–64 person-mrem) per container	Range of collective doses for four different location and storage scenarios.[2]
Waste Isolation Pilot Plant (WIPP)	0.02 person-mSv (2 person-mrem) per container	Waste is shipped to WIPP in RH72-B waste canisters (assume each canister contains 3 containers of 55 gallons). Waste handlers receive 80 percent of the dose.[3]

NOTE: Collective doses may not be representative of all RH-TRU waste generator sites. This table was compiled by the committee as an example of worker dose estimates on the basis of information gathered during the study. It is not intended to be complete and may contain inaccuracies.

[1] Memorandum from Dr. Rigley at Battelle Columbus Laboratories (BCL, 2000).
[2] Radiological risk assessment at Argonne National Laboratory-East (Cheng et al., 2002).
[3] Information presented to the committee during the second meeting (Wu, 2001).

The Battelle Columbus Laboratories in Ohio developed its own RH-TRU waste characterization plan, which involves visual examination of the waste during repackaging. The re-packaging step, which occurs in a hot cell, is necessary because the this site's RH-TRU waste containers do not meet transportation requirements. At this site, the worker dose per container (100 person-mrem per 5 persons corresponds to 20 mrem) is within the range of worker doses for CH-TRU waste characterization (Wu, 2001). It is unclear how representative Battelle Columbus Laboratories worker exposure data are for other RH-TRU waste generators. However, the Battelle data are informative because this is the only site actively characterizing RH-TRU waste.

A radiological risk assessment for RH-TRU waste has been performed for storage options at Argonne National Laboratory-East (Cheng et al. 2002). Although characterization of RH-TRU waste has not begun at this site, risk assessments have been conducted to analyze potential radiation exposures associated with various on-site storage scenarios. Detailed analyses of container handling activities at the Alpha Gamma Hot Cell Facility show that the major determinants of worker collective dose are manipulation of containers for waste loading and placing and securing container lids. These two activities account for 75 percent of the collective dose to workers (Cheng et al., 2002).

The WIPP facility has estimated worker doses for the handling and emplacement of RH-TRU waste (Wu, 2001). WIPP is currently not authorized to receive RH-TRU waste; estimates of worker doses are based on calculated average container surface dose rates, shielding characteristics of the RH72-B canister, and remote-handling of waste canisters. Worker doses are substantially lower than doses estimated at the generator sites primarily because of RH72-B canister shielding.

As shown in Table H.1, collective dose may vary widely between and within sites. For instance, container handling methods and procedures (the key determinant of worker dose) may differ. Further, the relative importance of worker exposures incurred during the characterization phase compared to that incurred in other handling operations will likely differ at the generator sites. Accordingly, it is important that ALARA programs be site-specific.

H.2 COMMITTEE'S EXAMPLE OF COST ESTIMATES

Characterization costs including facilities and instrumentation for RH TRU waste sampling and measurements are significant. According to DOE and the generator sites, the characterization of an RH-TRU waste container under the CH-TRU waste characterization requirements is estimated to cost in the range of $20,000 to $300,000 per container (Restrepo and Millard, 2001). In contrast, the cost of characterizing CH-TRU waste is about $3,800 per container (DOE-EM, 2001). The large range in RH-TRU characterization cost estimates is due to the variability in the waste content and the characterization processes among the sites. In addition, these cost estimates are not site-specific and take into account only a few characterization scenarios. Associated infrastructure costs in support of full characterization of RH-TRU waste have been estimated to be $100,000 per container (Restrepo and Millard, 2001). The Committee has not verified these cost estimates.

REFERENCES

BCL. 2000. Battelle Columbus Laboratory Memorandum from D.Ridgley. April 12.
Wu, C-F. 2001. TRU Waste Dosimetry. Presentation to the National Research Council Committee, 3–5 October. Albuquerque, N.Mex.
Cheng, J-J, H.Avci, D.Hecker, W.Bray, T.Bray, and C.Grandy. 2002. Radiological Risk Assessment for the Remote-Handled Transuranic Waste Storage Options at Argonne National Laboratory-East. Proceedings of the WM'02 Conference, February 24–28, Tucson, Ariz.
DOE-EM. 2001. Transuranic Waste Characterization Cost Analysis. U.S. Department of Energy-Office of Environmental Management. Memorandum from L.Wade, Director of the Waste Isolation Pilot Plant Office. February 23. Washington, D.C.
Restrepo, L.F. and J.B.Millard. 2001. A Risk/Cost Impact Analysis of Various Options for Characterizing Department of Energy Generated Remote-Handled Transuranic Waste. June 30. Omicron-01–012. Albuquerque, N.Mex. Presented in Notification of Proposed Change to the EPA's Waste Isolation Pilot Plant 40 CFR 194 Certification. July 16, 2001. Draft. Revision 1. Carlsbad, N.Mex.

Appendix I

Glossary

20.4.1	(The New Mexico Administrative Code Chapter 20, Section 4, Part 1): New Mexico Environment Department. Establishes the regulations for the management of hazardous waste consistent with the New Mexico Hazardous Waste Act and Federal Resource Conservation and Recovery Act (RCRA) regulations, Titles 40 Code of Federal Regulations (CFR) Parts 260 through 270.
20.4.1.200:	New Mexico Environment Department (NMED). This regulation (incorporating Title 40 CFR Parts 261.24, 261.31, and 361.33) requires the US. Department of Energy (DOE) to identify and list hazardous wastes.
20.4.1.500:	New Mexico Environment Department. This regulation (incorporating Title 40 CFR 264) requires DOE to conduct a detailed analysis of the hazardous waste components of transuranic mixed waste to obtain all the information on how to treat, store or dispose of the waste. DOE must demonstrate that the design and operation of the facility will minimize the possibility of the release of transuranic mixed waste, a fire, or an explosion. NMED prohibits the following at WIPP: 1. liquid waste; 2. pyrophoric materials; 3. non-mixed hazardous wastes; 4. chemically incompatible wastes; 5. explosives and compressed gases; 6. polychlorinated biphenyl concentrations; 7. ignitable, corrosive, and reactive waste; and 8. remote-handled transuranic mixed waste.
20.4.1.900:	New Mexico Environment Department. This document contains the hazardous waste permit program requirements issued by the NMED (incorporating 40 CFR 270). These requirements must be met by the DOE to receive NMED approval of the waste analysis plan submitted as Part B of the permit application (see Hazardous Waste Permit) for mixed transuranic waste.
10 CFR 20:	(Title 10 Code of Federal Regulations Part 20): United States Nuclear Regulatory Commission. Standards for Protection Against Radiation.
10 CFR 835:	U.S. Department of Energy. Occupation Radiation Protection. Establishes the standards, limits, and program requirements for protecting individuals from ionizing radiation resulting from DOE activities.

APPENDIX I

40 CFR 191: U.S. Environmental Protection Agency (EPA). Environmental Radiation Protection Standards for Management and Disposal of Spent Nuclear Fuel, High-Level and Transuranic Radioactive Wastes, Final Rule. December 20, 1993. Federal Register (FR) 58(242): 66398–66416. This regulation prescribes EPA environmental radiation protection standards that will apply to all sites (except Yucca Mountain) for the deep geologic disposal of highly radioactive waste. Congress required EPA to evaluate whether the WIPP complies with Subparts B and C of the disposal regulations set forth in this document for the management and disposal of transuranic radioactive wastes.

40 CFR 194: U.S. Environmental Protection Agency. Criteria for the Certification and Recertification of the Waste Isolation Pilot Plant's compliance with the 40 CFR Part 191 Disposal Regulations, Final Rule. May 18, 1998. Federal Register 63(95): 27353–2740. This regulation stipulates that DOE must provide a list to the EPA that identifies and describes waste characteristics that can impact the WIPP's performance. This list may be derived from methods that include process knowledge and non-destructive assay/examination. On May 18, 1998, the EPA issued a final rule certifying that the WIPP was compliant with applicable EPA TRU waste disposal regulations set forth in 40 CFR 191 and the compliance criteria of 40 CFR 194 (63 FR 27354).

40 CFR 194.22(b): U.S. Environmental Protection Agency. This section includes the quality assurance requirements for waste characterization activities and assumptions. The quality assurance provisions allow the characterization of waste by 1) peer review; 2) corroboration with new data, 3) confirmation by measurement, or 4) qualification of previous QA programs.

40 CFR 261: U.S. Environmental Protection Agency. Identification and Listing of Hazardous Waste. This part identifies those solid wastes that are subject to regulation as hazardous wastes under Parts 262–265, 268, 270, 271, and 124 of Title 40 of the Code of Federal Regulations. Codified in New Mexico as 20 NMAC 4.1, Subpart II.

40 CFR 264: U.S. Environmental Protection Agency. This part consists of "Standards for Owners and Operators of Hazardous Waste Treatment, Storage, and Disposal Facilities." This subpart establishes minimum national standards, that define the acceptable management of hazardous waste. Codified in New Mexico as 20 NMAC 4.1, Subpart V.

40 CFR 270: U.S. Environmental Protection Agency. This regulation establishes provisions for the Hazardous Waste Permitting Program under Subtitle C of the Resource Conservation and Recovery Act. This regulation and the associated State of New Mexico regulations require the permitting of the WIPP as a hazardous waste management unit. Codified in New Mexico as 20 NMAC 4.1, Subpart IX.

Acceptable Knowledge (AK): a term used by the EPA that encompasses process knowledge and results from previous testing, sampling, and analysis of waste. Acceptable knowledge includes information regarding the raw materials used in a process or operation, process description, products, and associated wastes. Acceptable knowledge documentation includes the site history and mission, site- specific processes or operations administrative building controls, and all previous and current activities that generate a specific waste.

APPENDIX I

ALARA (As Low As is Reasonably Achievable):	radiation protection program for minimizing personnel exposure to radiation. ALARA means making every reasonable effort to maintain exposures to radiation as far below the dose limits in the DOE guidance as is practical consistent with the purpose for which the activity is undertaken.
Audit:	a planned and documented investigative evaluation of an item or process to determine the adequacy and effectiveness as well as compliance with established procedures, instructions, drawings, and other applicable documents.
Buried Transuranic Waste:	radioactive waste meeting the current definition of TRU waste, which was disposed of by shallow land burial and other techniques at a number of sites owned and operated by the federal government in support of the nuclear weapons program from the 1940s through 1970. In 1970 the Atomic Energy Commission first identified TRU waste as a separated category of radioactive waste, and all TRU waste generated after 1970 has been segregated from low-level waste and placed in retrievable storage pending shipment to and disposal in an approved geologic repository. Most of this buried waste is considered irretrievable.
Compliance Certification Application (CCA):	DOE submits this application (Title 40 CFR Part 191, Compliance Certification Application for the Waste Isolation Pilot) to the EPA in order to request certification from the EPA for the WIPP facility.
Certified Waste:	containers of waste that meet the WIPP waste acceptance criteria.
Code of Federal Regulations (CFR):	1) a codification of the general and permanent rules published in the Federal Register by the department and agencies of the federal government. The CFR is divided into 50 titles that represent broad areas subject to federal regulation. It is issued quarterly and revised annually. 2) All federal regulations in force are published annually in codified form in the CFR.
Contact-handled (CH) waste:	Transuranic waste that has a measured radiation dose rate at the container surface of 200 millirem per hour or less and can be safely handled without special equipment when in closed containers. [LWA]
Cellulosics, Plastic, Rubber (CPR):	the characterization objectives for RH-TRU waste set forth by the EPA in 40 CFR 191 and 40 CFR 194 require that DOE account for the volume of CPR because of the potential gas generation from the decomposition of these organic materials.
Curies (Ci):	unit of radioactivity. One curie equals 3.7×10^{10} nuclear transformations per second. This unit reflects the intensity of a radioactive source.
Data Quality Objectives (DQO):	qualitative and quantitative statements that clarify program technical and quality objectives, define the appropriate type of data, and specify tolerable levels of potential decision errors that will be used as the basis for establishing the quality and quantity of data needed to support decisions.

APPENDIX I

Defense Waste: radioactive waste from any activity performed in whole or in part in support of DOE atomic energy defense activities; excludes waste under purview of the Nuclear Regulatory Commission or generated by the commercial nuclear power industry. It consists of nuclear waste derived mostly from the manufacturing of nuclear weapons, weapons-related research programs, the operation of naval reactors, and the decontamination of weapons production facilities.

Environmental Evaluation Group (EEG): the New Mexico Environmental Evaluation Group conducts an independent technical evaluation of the operations of the Waste Isolation Pilot Plant (WIPP) to ensure the protection of public health and safety, and the environment of New Mexico. The EEG has been serving New Mexico in this capacity since 1978. Public Law 100–456 articulates EEG's role and responsibilities relating to WIPP.

Gray (Gy): the standard unit of absorbed ionizing-radiation dose. One gray is equivalent to one joule of energy absorbed per kilogram of matter. One gray is equal to 100 rad.

Hazardous Constituent: those chemicals identified in Appendix VIII of 20 NMAC 4.1 Subpart II (40 CFR Part 261).

Hazardous Waste: as defined in 40 CFR 261.3, waste that because of its quantity, concentration, or physical, chemical or infectious characteristics, may cause or significantly contribute to an increase in mortality or an increase in serious irreversible, or incapacitating reversible illness, or pose a substantial present or potential hazard to human health or the environment when improperly treated, stored, transported, or disposed of, or otherwise managed. Hazardous wastes are listed in 20 NMAC 4.1 Subpart II (40 CFR Part 261) and/or exhibit one of the four characteristics—ignitability, corrosivity, reactivity, and toxicity in 20 NMAC 4.1 Subpart II (40 CFR Part 261).

Hazardous Waste Codes: numbers assigned to identify the EPA category of hazardous waste. Hazardous waste codes' assignment for RH TRU waste ensures that only those wastes that are permitted at the WIPP are disposed of and ensure waste compatibility during the operational phase at the WIPP.

Headspace Gas: the gas within the free volume at the top of a closed container (between the container lid and the waste inside the container) or containment, such as a drum or bin, containing TRU mixed or simulated waste. The gas may be generated from biological, chemical, or radiolytic processes; this includes contributions from volatile organic compounds (VOC) present in the waste.

Headspace Gas Analysis: headspace gas is sampled using a gas-tight syringe to draw a gas sample from beneath the drum or box lid. The sample is analyzed by gas chromatography and/or mass spectrometry for hydrogen, methane, and volatile organic compounds.

Land Withdrawal Act (LWA): Public Law 102–579 withdraws the land at the WIPP site from "entry, appropriation, and disposal." It transfers jurisdiction of the land from the secretary of the interior to the secretary of energy and reserves the land for activities associated with the development and operation of the WIPP. It requires DOE to comply with the EPA's radioactive waste standards and final disposal regulations and to conduct studies to analyze the impact of RH-TRU wastes on repository performance. It includes many other requirements and provisions pertaining to the protection of public health and the environment. The LWA was signed into law on October 30, 1992.

APPENDIX I

New Mexico Hazardous Waste Act (HWA):	the New Mexico legislation that establishes the state hazardous waste management program.
Newly Generated TRU Waste:	waste generated after the development, approval, and implementation of the TRU waste characterization program that meets the requirements outlined in the TRU waste characterization quality assurance program plan. Part of the inventory might not have been generated yet but is estimated to be generated at some time in the future by the TRU waste generator/storage sites.
Non-destructive Assay (NDA):	NDA is a general term for a number of techniques, such as gamma spectroscopy and passive/active neutron measurement. These techniques provide information on the radionuclide content of waste and sometimes on its spatial distribution inside containers.
Non-destructive Examination (NDE):	NDE is a general term for a number of techniques, such as radiography or computer tomography. Radiography is a non-destructive qualitative and semi-quantitative technique that involves X-ray scanning of waste containers to identify and verify waste container contents. Because of the shielding associated with RH TRU waste, computer tomography, which involves several sources to produce a three-dimensional image may be required rather than the more commonly used radiography.
Packaging:	the assembly of components necessary to ensure compliance with packaging requirements, it may consist of one or more receptacles, absorbent material, spacing structures, thermal insulation, radiation shielding, and devices for cooling or absorbing mechanical shocks.
Performance Assessment:	risk-based assessment of the safety performance of a nuclear waste facility. The purpose of the performance assessment for WIPP is to evaluate the ability of the repository to isolate radioactive waste from the accessible environment. The performance assessment organizes information relevant to long-term (i.e., over a 10,000-year period) repository behavior by assessing the probabilities and consequences of major scenarios by which radionuclides can be released to the environment surrounding the WIPP site. Important scenarios include those due to human activities, whether deliberate or unintentional, that might occur near the WIPP site and potentially compromise the integrity of the repository.
Pyrophoric:	spontaneously ignitable materials.
Process Knowledge:	the determination of waste container contents through the study of existing records on the production history of the waste.
Quality Assurance:	the planned and systematic actions necessary to provide adequate confidence that a structure, system, or component will perform satisfactorily in service.

Quality Assurance Program Plans (QAPP):	documents that describe the overall program plans and activities to meet the project's quality assurance goals.
Rad:	unit of absorbed dose. It represents 0.01 joules of energy absorbed per kilogram of matter. Rad and rem are important units regarding WIPP because requirements are expressed in rem, a derivative or rad.
Radioassay:	a term used to define measurement methods for determining the radionuclide content of waste and includes both non-destructive assay (NDA) and destructive assay (e.g. radiochemistry).
Radiological Survey:	Measurements of radioactive contamination levels or dose rates associated with a site together with the appropriate documentation and data evaluation. When AK indicates that some containers may approach 1000 rem/h or that some containers exceed 100 rem/h then radiological surveys of each container may be required. Industry standard survey instruments can be used in this process and are required to discriminate at 100 rem/h and 1000 rem/h.
Radiography:	a non-destructive, non-intrusive radiographic examination technique that enables a qualitative (and in some cases quantitative) evaluation of the contents of a waste container. Radiography utilizes X rays to inspect the contents of the waste container in real time. It is used to examine and verify the physical form of the waste for certain waste forms, identify individual waste components, and verify the absence of certain non-compliant items.
RCRA:	see Resources Conservation and Recovery Act.
RCRA Part B Permit:	issued for the WIPP on October 27, 1999 by the NMED. This permit (incorporating 20.4.1.500 and 20.4.1.900) determines that DOE's disposal plan for mixed TRU waste is acceptable. Also called the Hazardous Waste Facility Permit.
Rem (Roentgen Equivalent Man):	unit of absorbed radiation dose used to derive a quantity called equivalent dose. This relates the absorbed dose in human tissue to the effective biological damage of the radiation. Not all radiation has the same biological effect, even for the same amount of absorbed dose. Equivalent dose is often expressed in terms of thousandths of a rem, or mrem. The equivalent dose (rem) is determined by multiplying the absorbed dose (rad) by a quality factor (Q) that accounts for different biological effects caused by different radiations. Dose requirements regarding WIPP are expressed in this unit.
Remote-handled (RH) Waste:	transuranic wastes that have a measured radiation dose rate at the container survey of 200 millirem per hour or greater but not more than 1,000 rem per hour. This waste must be handled remotely (i.e., with machinery designed to shield the handler from radiation).
Resource Conservation and Recovery Act (RCRA):	established a system for tracking and regulating hazardous wastes from the time of their generation through disposal. The law requires safe and secure procedures to be used by hazardous waste generators in treating, handling, transporting, storing, and disposing of hazardous substances. RCRA is designed to prevent new uncontrolled hazardous waste sites. The law also regulates the disposal of solid waste that may not be considered hazardous. Note: 20 NMAC 4.1 and 40 CFR Parts 260–281 are the regulations for complying with RCRA with respect to hazardous waste and hazardous waste treatment, storage, and disposal facilities in New Mexico.

APPENDIX I

Retrievably Stored TRU Waste: waste generated after 1970. In 1970 the Atomic Energy Commission (predecessor to the DOE) first identified TRU waste as a separate category of radioactive waste. The same year, the Atomic Energy Commission determined that all TRU waste generated after 1970 must be segregated from low-level waste and placed in retrievable storage pending shipment to and disposal in an approved geologic repository. Federal facilities in Washington, Idaho, California, Colorado, New Mexico, Nevada, South Carolina, Ohio, Tennessee, and Illinois are currently storing TRU waste. See also Buried Transuranic Waste.

Roentgen: unit used to measure a quantity called exposure. The roentgen is that quantity of X or gamma radiation less than 3 MeV in energy that produces 1 electrostatic unit of charge, 2.58×10^{-4} coulombs, in 1 kilogram of dry air at 0 degree Celsius and at an atmospheric pressure of 760 mmHg. Many radiation measuring instruments measure the roentgen (ionization) directly. It is a measure of the ionizations of the molecules in a mass of air. The main advantage of this unit is that it is easy to measure directly, but it is limited because it is only for deposition in air, and only for gamma and X rays.

Safety Analysis Report (SAR): document representing a statement and commitment by the DOE that the WIPP can be operated safely and at acceptable risk. This document summarizes the safety analyses to ensure the safety of workers, the public, and the environment from the hazards posed by WIPP waste handling and emplacement operations during the disposal phase and hazards associated with the decommissioning and decontamination phase. The WIPP SAR is prepared to satisfy: (1) the commitments in the Working Agreement for Consultation and Cooperation between the State of New Mexico and the U.S. Department of Energy; and (2) to ensure compliance with DOE's 10 CFR 830 about nuclear safety management.

Sievert: unit of measurement of radiation dose equivalent. One sievert is the absorbed dose, expressed in gray, multiplied by a quality factor to account for different biological effects caused by different radiations.

Summary Category Group: categories for each waste stream to facilitate RCRA waste characterization that reflect the final waste forms acceptable for WIPP disposal. The waste summary categories are identified by the generators and are the following: S3000: Homogeneous solids S4000: Soils/gravel S5000: Debris waste

Transuranic Waste: radioactive waste consisting of radionuclides with atomic numbers greater than 92 in excess of agreed limits. A more precise definition, in DOE Order 5820.2A, EPA regulation 40 CFR 191, and the Land Withdrawal Act, is waste that is not high-level waste "contaminated with alpha-emitting radionuclides of atomic number greater than 92 and half-lives greater than 20 years in concentrations greater than 100 nanocuries per gram." The regulatory definition excludes actinide elements with atomic numbers between 90 and 92 (most significantly, Th and U isotopes), in agreement with the literal meaning of "transuranic." However, common usage of "transuranic waste" is often understood to include all actinides.

APPENDIX I

Transuranic mixed waste:	transuranic waste contaminated with hazardous constituents as identified in 20 NMAC 4.1 Subpart II (40 CFR 261), Subparts C and D.
Toxic Substances Control Act (TSCA):	Act enacted by Congress in 1976 to give EPA the ability to track the 75,000 industrial chemicals currently produced or imported into the United States. EPA repeatedly screens these chemicals and can require reporting or testing of those that may pose an environmental or human-health hazard. EPA can ban the manufacture and import of those chemicals that pose an unreasonable risk.
Visual Examination (VE):	process consisting of physically examining TRU waste by removing it from the container it was originally packaged in.
Volatile Organic Compounds (VOC):	RCRA regulated organic compounds that readily pass into the vapor state and are present in transuranic mixed waste.
Waste Acceptance Criteria (WAC):	set of conditions established for permitting transuranic wastes to be packaged, shipped, managed, and disposed of at the WIPP.
Waste Analysis Plan (WAP):	document describing the procedures that will be carried out at a facility to obtain chemical and physical analysis of each waste managed so that all information will be known to treat, store or dispose of the waste in accordance with 40 CFR 264.13.
Waste Characterization:	sampling, monitoring, and analysis activities to determine the nature of the waste.
Waste Matrix Code:	code assigned by the TRU waste generator/storage sites to categorize mixed and some non-mixed waste streams in the DOE system into a series of five-digit alphanumeric codes (e.g., S5400; Heterogeneous Debris) that represent different physical/chemical matrices. These codes were developed by DOE in response to the Federal Facility Compliance Act of 1992.
Waste Profile Form:	form that waste generator must complete to properly identify and document the characterization of any solid, liquid, hazardous, radioactive, or mixed waste. The Waste Profile Form must provide a complete and concise description of the waste, including the details of the generating process. The Waste Profile Form process provides generators with guidance to help make the determination of the waste's physical, chemical, and radiological characteristics with sufficient accuracy to permit proper segregation, treatment, and disposal according to the final treatment/disposal facility's waste acceptance criteria.
Waste Stream:	waste material generated from a single process or activity or as multiple containers with similar physical, chemical, or radiological characteristics.

Appendix J

Acronyms

AK:	acceptable knowledge
ALARA:	as low as reasonably achievable
CAO:	Carlsbad Area Office
CBFO:	Carlsbad Field Office
CH-TRU:	contact-handled transuranic
DOE:	U.S. Department of Energy
EPA:	U.S. Environmental Protection Agency
NMED:	New Mexico Environment Department
PUREX:	plutonium uranium extraction
RH-TRU:	remote-handled transuranic
RCRA:	Resource Conservation and Recovery Act
SCG:	summary category group
WIPP:	Waste Isolation Pilot Plant
WWIS:	WIPP Waste Information System